# JACK COSTA

Dario De Bortoli

# JACK COSTA

Dario De Bortoli

Translated from the Italian
by Sara Jane Pearcey

Autonomedia

ISBN: 978-1-57027-288-2



Autonomedia
POB 568 Williamsburgh Station
Brooklyn, NY 11211-0568 USA

info@autonomedia.org
www.autonomedia.org

Contents

*Photographs follow page 74*

# Preface

It was back in the early eighties that I first heard of Jack Costa, when my acquaintance with Mariarosa Dalla Costa had grown into a firm friendship, through which I learned about her main family events, in particular the singular story of her grandfather, Giovanni, who emigrated from the Veneto region in search of gold in Alaska.

Gradually other experiences sparked my curiosity. The first was a visit to the small cemetery in Pederobba where, summed up on a simple marble tombstone, I read about an extraordinary existence. Then again my interest was fired by the stories of legendary happenings told by Mariarosa, and her mother, Maria Ghidelli, her brother and sister, Gianni and Franca, and other relatives, whom I gradually got to know; this made me want to read the "papers" that Mariarosa had carefully guarded in a few files. Actually, her collection was quite abundant, from various articles in newspapers and journals, written by the journalist Lino Pellegrini, to accounts of the "interviews" made over time by Franca to her aunt Rita, and by Mariarosa to her father, Francesco; from Maria Ghidelli's notes recording her recollections of her husband and mother-in-law, to some translated fragments of Alaskan publications on the gold-digging enterprise on ice undertaken by Jack and his brother, Francesco Dalla Costa, who came to be called Frank Costa. I felt it would be worthwhile to bring this story to light, to piece the fragments together, and tell it all in a book. The basic material was to hand, but I still had to set this story of initiative and courage, of great fortune and tremendous misfortune, into its developing historical and social context. Three fields of investigation were required for me to reconstruct the backcloth for the events that took place: the conditions of peasants in the second half of the nineteenth century, their emigration to various countries, and the gold rush. For the first two, I could rely on the exhaustive literature available, but for the third, which was more tricky, particularly concerning the Alaskan context, the research was to be more painstaking. Yet another troublesome line of investigation was needed in order to complete the picture: reconstructing the paths and vicissitudes of other members of Giovanni Dalla

Costa's family, in particular his brothers, Francesco, Gaspare and Giacomo. The task of finding out about the life of Francesco Dalla Costa, who had followed Jack in his Alaskan adventure, was relatively straightforward, since one of his descendants lived in Italy. Concerning the other two brothers, whose only relatives I knew of were living in Brazil and France, I had to play it by ear and proceed tentatively.

Helped by Mariarosa, I dispatched letters to Brazil - to addresses that were at least twenty-years old; to trace relatives in France, an email message was sent straight to the Local Council of Figanières asking whether details on, and the whereabouts of, Giacomo Dalla Costa's descendants were available. For a long time nobody replied, but then, first from France and later from Brazil, the people we were looking for took on bodies and voices. From Figanières it was Gilberte Arnoux, Giacomo's granddaughter, who wrote, sending us recollections, letters and photos with the help of her daughter, Colette, and her son-in-law, Jacques Escaillas, and also her grandsons, Eric and Fabien. From Guaporè, in Rio Grande do Sul, Marly Dalla Costa, Gaspare's granddaughter, was the one to write and, in turn, send photographs and documents including an extremely important one: the passport of Luigi, father of the two gold-digger brothers. The document authorized Luigi, together with some of his family members, to leave Genoa for Montevideo on March 16, 1890. We now had an exact date, and the plot was beginning to unravel: from here to the other side of the ocean.

Some time later, more findings arrived, including a photograph from Ivan Luiz Dalla Costa, Gaspare's grandson who lived in Caxias Do Sul; it showed the very first gold nugget extracted in Alaska by the two gold-digger brothers. The emotion and enthusiasm expressed from across the seas by relatives, together with their wish to collaborate and make a contribution to the book that could give them back a story that had almost been lost, spurred my determination. Now I had to start writing, and as soon as possible. So, fresh with the new documentation that had arrived from Brazil and France, and from research undertaken at the state, county and parish archives, port authorities, libraries in Alaska and websites on the Internet, I drew up the pages that follow, integrating them with the news that I received during meetings in Pistoia with Frank Costa's grandson, Francesco Bonfiglioli, and his wife, Emanuela, and in Cornuda, with Jack's grandson, Wally Bresolin. The information that Remy and Rosmary Bresolin, other grandchildren of Jack's, sent me from Australia, where they had emigrated to some time back, was also precious. Finally, an important milestone was the meeting held with the journalist and actor, Giorgio Comaschi, and the researcher, Claudio Busi, both of whom are in Bologna working on a novel inspired by the life of Felix Pedro, a companion in the work and adventures of the Dalla Costa brothers. All that I have written in this book is therefore founded on documents and recollections, the story being as authentic as it is extraordinary. It is the true tale of a family which, in the waning nineteenth century, after enjoying centuries of stability in a small medieval borough, was suddenly forced to disperse throughout the world, some in search of bread and some for fortune. And it is the story of a man who, one hundred years after his quest, still offers us exceptional testimony to human determination in searching for, whatever the cost, the chance of a better life.

## Acknowledgments

This book was written thanks to the cooperation of scholars, and other kind people, all of whom in their different ways helped me gather documents and information. First, Professor Terrence Cole of the History Department of the University of Alaska Fairbanks, who, several years before had provided the descendants with news and documents from the Alaskan records on Jack and Frank Costa. The journalist, Jean Murray, alongside her husband, Robert, and daughter, Jenny, residing in Anderson, Alaska, were go-betweens for these dispatches: they had come to Italy to see the descendants of Jack Costa, after a chance meeting, during a flight to Alaska, with one of his daughters, Rita, who was traveling on the tracks of her father. Particular thanks go to Jean for her archive research activity at the University of Fairbanks. Precious help was also given by Rose Speranza, a librarian at the same University. Of fundamental importance, moreover, were the contributions from Professor Gianni Penzo, in charge of the General Archive of the University of Padua, and teacher of archiving, for his indications concerning sources. Heartfelt thanks also go to Professor Luigi Contegiacomo, Director of the State Archives of Rovigo for his numerous suggestions. The Pederobba members of the association "Treviso citizens in the world" steadily cooperated in the research in many ways; their keen involvement and support encouraged us to continue with this work. In particular, help came from the Association's Chairman, Angelo Bresolin, and his wife, Rosa, together with their daughter Bertilla and Beniamino Bresolin, who conducted diverse archive research activities. The excellent translations of letters in Portuguese by Regina Moretto, and the photographic-digitalization provided by Alessandro Cemolin were also very much appreciated. Thanks also go to our friends, Francesco and Loredana Pedini, Cristina Zuppel, Adamantonio Carriero, Salvatore Italiano and Elisabetta Mozzo for their painstaking revisions and thoughtful suggestions.

# 1
# A Happy Family in a Veneto in Crisis

Giovanni Dalla Costa was born on April 21, 1868, only two years after Veneto had been freed from Austrian rule and annexed to Italy. During his childhood, he would never have known about the story that started some twenty years back, thousands of kilometers away in California, a phenomenon that, more than any others, was to shape his life. He is hardly likely to have heard about it in his early years, spent in his birthplace, the hamlet called Costa,[1] or Alla Costa, as referred to in the Austrian register of 1842, consisting of a few poor peasant houses huddled together on the slopes of Monfenera, a few hundred meters above the village of Pederobba, at the foot of Monte Grappa.

Giovanni's family members were nicknamed "i pomer" because they grew abundant quantities of apples ["pomi" in Veneto dialect] on strong, healthy trees that, situated on the sunny south-facing slope, yielded some of the most tasty fruit in the area. Small tenant farmers, they also worked for a few people here and there, including the signori Damini, a family of jewelers, the Fabbriceria in the parish of Onigo, the Mensa Capitolare in the Cathedral of San Pietro at Treviso, and the Oratory of San Sebastiano at Pederobba.[2]

While he was growing up, Giovanni passed his time with a splendid landscape right before his eyes: ahead were the wavy profiles of the nearby Asolo hills and, a little to the left, the winding path of the river Piave — just before it opened out onto the Veneto plain. His childhood days, in harmony with the rhythm of the seasons, were spent helping his father Luigi and mother Teresa, and working the fields together with his brothers. Giovanni had several siblings: Maria Anna, born five years before him; Francesco, a couple of years older than him; then came Giacomo, Elisabetta and Gaspare, born two, six and eleven years, respectively, after him.[3] With them at Easter, as a child, Giovanni played *tirar ai ovi coi zinque schei*[4] and then the brothers and sisters came together to *"magnar la fugassa."*[5] On January 20, at the festival of San

Bastian, the children nibbled peanuts, *straccaganasse* [dried chestnuts] and *carobole* [carobs]. Every Sunday, once mass was over, they unfailingly queued up in front of Sborel's stall, which would be bursting with colored mints, *putineti* [rag-dolls], *trombete* [small trumpets] and *baloneti* [small balloons].

For a while, Giovanni Dalla Costa also went to school. Probably he only attended the first few years of elementary school, where he learned to read rather than write, although he mastered writing a small d, which was to remain his signature throughout his lifetime; this was enough for him.[6]

Giovanni had a peaceful childhood. Although his family paid rent to farm other people's property, he enjoyed a certain well-being, due, above all, to the fact that both his father, Luigi, and his mother, Teresa Morgantin, were left as only-children, which was rare in Veneto in those days, but meant that the two spouses enjoyed the entire estate of their two respective families, and that their small inheritances enabled them to have a better lifestyle than other families with similar backgrounds. Luigi, born in 1834, was the son of Anna Negrin, the second wife of his widowed father, Francesco; and he first saw the light when his father was already over forty years of age. In 1819, Francesco's first wife, Maria Baccin, had conceived another male child, Giovanni, who was stillborn. Teresa, born in 1838, was tall and blond; she had four sisters and a brother, but four of her siblings died in childhood, leaving one, Corona, who lived for 16 years, from 1841 to 1857. So, when Teresa married Luigi in 1861,[7] she brought all her family's possessions with her. In her new family home, she enjoyed dressing up in jewelry on special occasions, and used to cook with beautiful copper saucepans; she also had a sum of money deposited in the bank of Montebelluna.[8] The Dalla Costas owned working equipment, some farm animals and, thanks to their farming activities, found it relatively easily to provide their six children with food, clothing, some education, and the odd treat every now and then.

Yet the general situation in the Veneto region was somewhat different. After the Habsburg domination of the Lombardy-Veneto area, in 1859 Lombardy became Italian, while the Veneto stayed under Austrian rule for another seven years until the plebiscite of 1866, when it was annexed to Italy. During this seven-year period, the region suffered a harsh economic backlash, falling prey to a situation which, on the one hand, hindered its long-standing exchanges with the Lombardy merchants and, on the other, meant that the Habsburg Empire had come to consider it an appendage: since it was about to be detached, it was no longer considered worthy of investments; yet, although a commercial non-entity, it was constantly burdened by increasingly heavy taxation.

Once Veneto was annexed to Italy, the situation went from bad to worse. The few remaining trade links with Austria disappeared altogether and Veneto, which had remained backward, uncompetitive and, therefore, devoid of openings on the market to the other regions in Italy and abroad, came to be isolated and had to struggle to survive.[9] As well as suffering from unfavorable geopolitical contingent factors, the Veneto of those years lacked the structural bases of crucial importance in keeping up with the times. Agriculture, almost the only activity in the region, was becoming seriously underdeveloped. Property was over-fragmented, and this precluded the accumulation of enough capital to invest in new technologies, and any specialization in crop production was an unfeasible option for the individual land-worker, since such a choice would have exposed him to a precarious market that could not guarantee adequate economic returns, especially in the early eighties, when the importation of highly competitive crops from abroad was on the increase; the thriving trade in American, Russian, Hungarian and Danubian grain, Indian rice, cocoons and raw Chinese and Japanese silks crushed the Veneto farmers, who provided the very same products, but used obsolete technology to produce them.[10]

The tenant farmer, on his part, had to cultivate a bit of everything to try and ensure for himself at least a modest livelihood. Wheat, wine and silk-cocoons to pay the rent, and maize for the family to eat. Fodder was in short supply, crop rotation was reduced to a minimum, and livestock, almost always inadequate for the work to be done and the land to be fertilized, were often used as draft animals rather than for meat, milk and dairy produce.

## Notes

1 The hamlet, situated in an area called Alla Costa in the Austrian land register of 1842, is still called Costa Alta by its inhabitants; now, however, its official name is simply "Costa."

2 State Archives of Treviso — Austrian land register of 1842.

3 Luigi and Teresa had two other children: Maria Anna, their first daughter, who died in 1862, only 23 days after her birth, and Gaspare D., who was born in 1873 and died seven days later (Parish Archives of Pederobba).

[4] This game consisted of throwing a coin in turns onto a shelled hard-boiled egg. The winner was the one whose coin stuck in the egg first.

[5] Eating *"focaccia,"* which is flat, oven-baked Italian cake.

[6] These recollections of Giovanni's regarding his childhood were taken down by Giovanna Franca Dalla Costa, daughter of Francesco, son of Jack Costa, during her talks with her aunt, Rita (Margherita), daughter of Giovanni.

[7] The information on the genealogy of the families of Luigi Dalla Costa and Teresa Morgantin was taken from the Parish Archive of Pederobba.

[8] Letter from Gilberte Arnoux, daughter of Clotilde, daughter of Giacomo, brother of Jack Costa; Figanières, dated December 30, 2004 and sent to Mariarosa Dalla Costa, daughter of Francesco, the son of Jack Costa; in it, Gilberte recounts her grandfather's recollections. Thanks to her contribution, we managed to reconstruct the story of Giacomo and obtain important information on Luigi and Teresa, and, above all, discovered that the family was not poverty-stricken — as had been claimed by journalists writing about this issue.

[9] A. Lazzarini. *Campagne venete ed emigrazione di massa (1866–1900)* [Veneto Countryside and Mass Emigration (1866–1900)], Istituto per le ricerche di storia sociale e di storia religiosa, Vicenza, 1981, p. 101.

[10] A. Lazzarini, *op. cit.*, p. 193.

# 2
# Misery and Privileges

Society, as stagnant as the economy, was held back by the alliance between the land-owning aristocracy, the landed bourgeoisie and the church, all three social classes intent on keeping their privileges intact. But what exactly were these privileges?

In order to answer the above question, it is worthwhile to bear in mind that in those times in Veneto, agricultural workers fell into three fundamental categories. Above all, there were the farmhands, some of whom had a fixed job, although most were seasonal workers. In the nineteenth century, starting in the seventies, and even more so throughout the eighties, these workers had to bear the brunt of two emerging problems: on the one hand, competition from tenant farmers who, in ever increasing numbers, were forced to look for temporary work, since their rental contracts were no longer renewed, and, on the other, the decreased demand for labor after the arrival of large agricultural machinery — first the threshers, followed by harvesters and mowers. Then there were the share-croppers, bound by a contract that normally lasted a year yet tended to be tacitly renewed. Now, this kind of contract was increasingly often terminated by the owner, who, to maintain or boost his income, became more and more demanding, year by year. The share-croppers, often deprived of their reserves, raised animals leased by the owner, and seldom owned the equipment required for their activities. The entire produce was delivered to the owner, who valued it at prices current at harvesting time and, on that basis, split his earnings with the share-cropper. But the owner could then sell the produce when he considered it most advantageous, that is when prices were more favorable, thus making a greater profit, from which the share cropper was excluded. The contract also included the provision of a certain quantity of gratuities, eggs, fowl, meat, butter, fruit, grapes and wood, as well as the obligation to help out on the owner's estate. And finally there were the tenants, the category to which Luigi Dalla Costa's family belonged, who were allowed to manage small farms in exchange for a fee, either in kind or in cash. Yet rents, which were already high, continued to increase, not only due to the fierce competition between peasants searching for land to work on, but also due to the tendency of the owner to make as much as he could out of his land so as to keep up an income that was constantly threatened by the increase in taxes and the falling prices of agricultural goods.

In Veneto, in the wake of the unification, rental agreements for payment in kind or mixed (i.e. payment in kind and in cash) were increasingly often terminated for two reasons: one was that the tenant farmers tended to cultivate more and more maize for their own families and to produce less and less wine and silkworm cocoons for the owner, and the other was that the owners wanted to protect themselves from the crisis, with its fluctuating prices. So, in the seventies and eighties, increasing numbers of owners demanded rent in cash rather than in goods. Land-workers were thus forced to scrape money together by directly selling their produce, hampered by their extremely weak position, and struggle against an unpredictable market, characterized by strong fluctuations in prices, while buyers were ever more ready to exploit the moments of harvest, when goods were in greater supply, thus triggering a sharp drop in prices.

These were the conditions of peasants, and the privileges of landowners. The fierce determination of the latter in wanting to keep their dominant position intact was one of the main causes of the progressive and relentless crisis that Veneto agriculture plummeted into, especially from 1876 onwards. Other contributory factors were the violent hailstorms in the months of June and July of the same year, and the summer drought, made worse by the lack of adequate irrigation systems. The harvests were getting poorer, and cereals for men and fodder for animals scarcer, with a consequent reduction in cattle. The winter frosts led to famine, and the peasants were at the mercy of the harsh climate and relentless taxation. Thus the living standards of the rural population progressively deteriorated, their diet becoming increasingly poor. This adversity was affecting an ever larger section of small landowners, tenants and farm hands. The growing dependence on money, the restrictions imposed on rights for grazing and gathering firewood, the break-down of home spinning and weaving, together with the progressive disappearance of temporary work that, until then, had enabled families to make ends meet, caused the farmers to pull up their vines, cut down their fruit trees and simply grow cereals, limit crop rotation and impoverish the land. In these conditions anything unforeseen became a lethal tragedy for him who, already in dire economic straits and burdened with debt, was no longer able to set anything aside, not even in the more favorable years.

## Notes

[1] A. Lazzarini, *op. cit.*, p. 79.

[2] A contract by which the two parties agree to breed animals. The owner provided the animals, and the small farmer did the work.

[3] A. Lazzarini, *op. cit.*, p.p. 123–124.

[4] A. Lazzarini, *op. cit.*, p. 53.

[5] L. Magliaretta, "Alimentazione, casa, salute," [Diet, home, health], in S. Lanaro (ed.), *Il Veneto*, Giulio Einaudi, Turin, 1984, pp. 648–651.

# 3
# Maize, Taxation and Disease

Between the seventeenth and eighteenth centuries, maize was present in the agriculture of the Veneto countryside on a massive scale. It was thanks to this cereal that, on the one hand, the population had grown, but, on the other, maize played a role in making regional agriculture backward by supplanting millet, which was more profitable, mainly because it could be exported to foreign markets; maize also replaced oat, which had been valuable animal fodder. Moreover, the increasingly large land surface taken over for maize production led to the depletion of woodland, grazing land and pastureland, thus triggering a sharp drop in cattle and sheep rearing. Although a long-standing phenomenon, the tendency to cultivate maize revealed its dark side starting from the second half of the nineteenth century. Once all the other farm produce had been sold for cash, the Veneto peasant family's diet consisted almost exclusively of maize, which merely alleviated their hunger.[1] In those times, the daily diet of most peasants consisted of a few hectograms of *polenta* [corn-meal], on rare occasions combined with a little salted fish. Sometimes soup was brewed with beans and lard; meat, extremely scarce, almost always consisted of scraps. Milk and dairy produce, eggs, chicken and rabbit, if and when available, were increasingly often exchanged for maize flour, for quantity had precedence over quality.[2]

This grim setting was exacerbated by the onset of diseases — affecting both plants and people. Grapevines were attacked by mildew, rot, plant-lice and mould. Silkworm breeding had been in a state of crisis since the early fifties due to disease and larval flaccidity. Cattle died of carbuncle and foot and mouth disease. Humans, on the other hand, who fed almost exclusively on maize, fell sick above all with *pellagra,*[3] but endemic diseases, including malaria, diphtheria, smallpox and typhus, were also decimating them. Between 1884 and 1886 a cholera epidemic killed thousands.[4]

On another front, Veneto was one of the regions to be most heavily burdened not only by direct taxation from the Kingdom of Italy, but also by an enormous accumulation of surtax, both from the province and the local municipality. In 1881, for every 100 lire of land revenue tax, surtax amounted to more than 140 lire, the highest proportion in the Kingdom. The increases were unbearable: there was a tax on the family, one on farm animals, others on pig slaughtering and on salt, making it difficult to buy even the small quantities needed to flavor polenta. On top of that came the notorious tax on polenta, the so-called tax on hunger. In 1868, the year of Giovanni Dalla Costa's birth, the loathed tax on milling was introduced: as from January of the following year, the tax on milling cereals amounted to 2 lire for every 100 kilos of wheat, 1.20 lire for oats, 0.80 lire for corn and rye and 0.50 lire for other cereals and chestnuts. The tax levied on milling, a scourge on peasants, had been impoverishing them for fifteen years. After an initial reduction, decided upon between 1879 and 1880, it was at last eliminated only on January 1, 1884. Introduced to cure the sick finances of the state, it drained the meager reserves of families, depriving them of any left-over cash needed for paying rent, buying clothes and equipment, and enjoying a more varied and wholesome diet. Millers, charged with the collection of dues, asked for almost a quarter of the corn brought for milling in lieu of tax, and as a fee for their services.[5] Failure to pay milling tax could lead to the confiscation of the small properties.[6]

On top of this terrible increase in taxes people had to continue to pay a tenth part of their income for the feudal tithe, a contribution toward the costs of worship and the maintenance of the clergy, which parish priests and chaplains demanded ever more aggressively and despotically as resources became scarcer for all and sundry.[7]

The peasants found it hard to live in the midst of this hardship, with the continuous need for money to pay their dues, increasingly high rents, the expenses incurred for conducting their activities, and paying for flour when their harvest was inadequate. Mortgage loaning was widespread, brokers were spewing out tiny pawns, and peasants were forced to resort to usurers because banks, although popular, stipulated impossible conditions; demanding guarantees that most peasants could not give them, and conceding loans with early deadlines, of from four to six months, which was far too exacting for agricultural rhythms; they also demanded interest payments as high as 9% and were relentless at the moment of exaction, ruling out any possibility of making re-payments in installments.

Thus tenant farmers were more vulnerable to credit institutes. Soon they had to sell part of their reserves to pay for the interest on their loans and they often ended up in an untenable position, and had to give up, and terminate their rental agreements. After this, few options were left open to them: either they competed with others who, like them, were looking for new land to rent in increasingly unfavorable conditions, or they worked as farm-hands; the last option open to them was to emigrate.

## NOTES

1 A Lazzarini, *op. cit.,* p. 96.

2 A Lazzarini, *op. cit.,* note 38, p. 660.

3 " . . . the most widespread disease was pellagra. The first signs of this terrible condition, a form of generalized anemia, appeared on the exposed parts of the body, which took on a bluish hue; then the gastrointestinal tract was affected, with vomiting and diarrhea, while the tongue became swollen. Simultaneous to this, there were disturbances of the nervous system, with depression and confusion. Finally, those with a more acute form of the disease went insane. They hanged themselves or, terrified, threw themselves into a torrent…." (*La valigia dell'emigrante* [The emigrant's suitcase], La Valigia, Vicenza, 1999, p. 21). The epidemic, caused by a severe lack of, and imbalance in, diet, lasted well beyond the period analyzed in this book. It was thanks to Doctor Giovanni Lino Ferrari, a generous medical practitioner from Fossalta di Piave, that studies on the curative effects of nicotinic acid were conducted. In 1938, alongside Prof. Gino Frontali, he published (Minerva Medica) an article entitled "Le forme nervose della pellagra curate con acido nicotinico" ["Nervous forms of pellagra treated with nicotinic acid"], this approach was remarkably successful, prompting several other studies of its kind.

4 A Lazzarini, *op.cit.,* pp. 55–57. On the general and health conditions of the Veneto population in those times, see the recent work by G. Dalla Zuanna, A. Rosina and F. Rossi (eds), *Il Veneto. Storia della popolazione dalla caduta di Venezia a oggi* [The Veneto. History of the population from the fall of Venice to present times], Ricerche Marsilio, Fondazione Nord Est, Marsilio, Venezia, 2004.

5 A Lazzarini, *op.cit.,* pp. 166–175.

6 "Between 1875 and 1881, more than 61,831 small holdings were confiscated. Between 1884 and 1901, the number of properties taken from small farmers had risen to 215,759; agricultural court sales for debt incurred by private

owners amounted to 70,774 between 1886 and 1900" from E. Sereni, *Il capitalismo nelle campagne (1860–1900)* [Capitalism in the Countryside (1860–1890)], Einaudi, Turin, 1956, p. 279 ff, quoted by G. Erler, *L'emigrazione italiana nello Stato di Rio Grande do Sul (Brasile) tra il 1875 ed il 1914* [Italian emigration to the State of Rio Grande do Sul (Brazil) from 1875 to 1914], in R. De Felice (ed.), *Cenni storici sulla emigrazione italiana nelle Americhe e in Australia* [Historical outline of Italian Emigration to the Americas and Australia], FrancoAngeli, Milan, 1979, note 5, p. 62.

[7] A Lazzarini, *op. cit.,* pp. 173–175.

# 4
# Giovanni as a Teenager

While growing up in this difficult world that was fraught with further difficulties, Giovanni could, of course, count on the security that came from living in a family that, all in all, managed to grapple with the mounting social and economic problems. He worked in the fields and, together with his brothers Francesco and Giacomo, took the cattle to graze in the fields near the hill — where there were plenty of apple trees. Distracted by other thoughts, the boys often lost sight of the animals, which were left free to eat this fruit and at night would moan because of the inevitable symptoms.[1] "Weird disorders," the three accomplices would explain, tongue in cheek.

As the years went by, Giovanni grew more vigorous, healthy and strong, and started to gain a better understanding of what was happening around him. In the evenings, as they gathered together and chatted in the warm cowsheds, but also among friends, conversations would turn to the problems of one family or another, the disease that had stricken a fellow countryman, animals or crops, and also to emigration, the drastic choice an ever increasing number of individuals and families were now being forced to make.

But Giovanni's entire family was well and healthy. In 1886, his brother Francesco started on his three-year stint of military service in the Third Genio[2] Regiment while his sister, Maria Anna, who was expecting her second child, Francesco Andrea, had at last married Fortunato Viviani by whom, two years before, she had conceived her daughter, Santa Teresa. With his easygoing, jovial character, Giovanni made the most he could of his youth, and, whenever he had some free time, played a hand of cards, drank a glass of wine, and chatted with his friends. But in the evenings, one topic continued to rear its head increasingly often: emigration.

And certainly in the Veneto setting of those times, emigration was sharply and steadily increasing, hand in hand with the overall impoverishment, the tax burden, diseases, debts, and growing desperation. There were two kinds of emigration. One was temporary, large contingents of Veneto peasants leaving their land every year for a few months, traveling to central and Western European countries. The second half of the nineteenth century saw this long-practiced short-term emigration mushroom.[3] While around the year 1880, this trend involved about 30,000 individuals, mainly men, small farm owners, who went abroad to find another job to supplement the meager income from their usual activities, throughout the decade from 1887 to 1897, this option was taken every year by about 65,000 Veneto peasants. The flow increased and, by the end of the century, the number had gone up to over 100,000. The destination of half of these workers was Austro-Hungary, a quarter went to Germany, and the remainder to Switzerland, France and the Balkans.[4]

Immediately after the annexation of Veneto to the Kingdom of Italy, this so-called seasonal emigration was followed by permanent emigration. According to official statistics, in the decade from 1876 to 1885, about 50,000 people left Veneto and, in the following 15 years (1886–1900), another 350,000 departed. This massive exodus involved one out of every ten inhabitants in the Veneto, thus making the region the one in Italy with the highest number of emigrants. But where did these people go? Eighty-two per cent went to Brazil.[5] Why this particular country? Because there, even if already in 1875 a policy for attracting Italian labor had been started, in 1888 the chains that for centuries had bound black slaves deported from Africa were finally broken. And from that moment on, Brazil, the last country to abolish slavery, was determined to replace the work done by slaves with that done by European emigrants, including Italians, more often than not from Veneto. Brazil was thus following in the footsteps of the United States where, as from 1865, slaves had been replaced by immigrant workers (in 1880 in Louisiana alone, 150,000 Italians worked on the cotton and sugar-cane plantations; many were women and children, whose conditions were similar to those of slaves, the most trivial misdemeanor being punished by a ferocious beating from the owner). Brazilian masters who, thanks to the arrival of this new labor-force could carry on with their activity unimpeded, were free to continue to treat these workers just as they had always treated their slaves. Rich and powerful, they controlled Government and Parliament, formed the backbone of the economy and, were the law unto themselves; but they greatly needed men and women who could work the land, preferably within the framework of a family unit, since this way they were more dependent and easier to blackmail — more ready to knuckle down to the working conditions.[6]

However, the beginning of the great Veneto emigration to Brazil dates back to a full 13 years before the end of slavery, to 1875, when Caetano Pinto Junior, the famous Brazilian emigration entrepreneur, stipulated with the Emperor a contract for the recruitment of 100,000 settlers in the space of ten years: Italians, Spaniards, Austrians and Swedes, of from two to forty years of age. Adults had to be experienced farmers, hard-working and of sound moral fiber, which is exactly what the Veneto peasants were.[7] Brazil's immigration program was mainly designed to meet the needs of huge agricultural estates, especially coffee plantations. The landowners of S. Paolo, Rio de Janeiro, Minas Gerais and Espirito Santo, firmly intent on securing the supply of farmhands, above all lobbied for immigration legislation that favored the cultivation of large estates, and they did their utmost to secure a complete monopoly over the available work-force. While before 1888 the work of immigrants was integrated with that of slaves, in the ensuing years, as slaves were freed and mass emigration occurred, new immigrants were expected to directly replace the lost labor force. And they worked so well that coffee plantations became the mainstay of Brazil's economy, coffee soon becoming Brazil's leading export, worldwide. However, not all immigrants ended up as farm laborers or plantation-workers given a fixed wage or paid on the basis of piece-work. Unlike the provincial governments, whose policies were designed to satisfy the interests of the large landowners, the Emperor of Brazil wanted to create farms and small colonial centers based on land ownership and on diversified cultivation so as first to feed the working unit, and then supply the market. But this policy could be carried out only in those areas in which coffee was not grown, mainly central and southern provinces, such as Paranà, S. Caterina and Rio Grande do Sul. The plan took root, in particular, in the area of Rio Grande, where the imperial colonization policy had two main objectives: one, economic and social, was to populate and utilize barren land, and the other, strategic, was to protect a geographic area of crucial importance. In fact, if the political and economic life of the Empire was concentrated in the province of the center, the south, very fertile but completely deserted and difficult to control due to the scarcity of roads and inhabitants, was at risk of attacks from bordering states. Colonization would address these issues by creating a network that was not only agricultural, but also civil, with residential centers, administrative organs and roads. Before the seventies, the immigration of Germans and other ethnicities had been encouraged in these lands. German immigrants in particular were well organized and self-sufficient and played an important role in public life, but they made up a community that kept itself to itself, separate from the rest of the population. What was needed therefore was an influx to the south of Brazil of immigrants eager to become integrated in their new country. This is why, in the period from 1875 to 1894, the average number of people who emigrated from Veneto to Brazil was constant and extremely high, higher than that from any other region in Italy. During the three-year period from 1888 to 1891, in particular, the exodus of Veneto emigrants towards the Brazilian lands exceeded the exceptionally high total of 70,000.[8]

## Notes

[1] Letter from Gilberte Arnoux to Mariarosa Dalla Costa, Figanières, February 26, 2005, in which she relates the recollections of her grandfather, Giacomo, brother of Giovanni Dalla Costa.

[2] Archive of the Local Authority of Pederobba.

[3] A. Lazzarini, *op. cit.*, p. 26.

[4] *La valigia dell'emigrante* [The emigrant's suitcase], *prev. cit.*, pp. 47, 60–61.

[5] A. Lazzarini, *op. cit.*, pp. 27, 189, 221–222, 226.

[6] *La valigia dell'emigrante, prev. cit.*, p.48.

[7] P. Brunello, *Emigranti* [Emigrants], in S. Lanaro (ed.), *op. cit.*, pp. 581–582.

[8] Regarding the two Brazilian immigration policies, and for a more general picture, see G. Erler, *op. cit.*

# 5
# The Agents' Promises

A new figure to appear on the Veneto horizon in those tough times, soon to become a ubiquitous specialist in the traffic of men, was the emigration agent, whose task it was to encourage impoverished peasants to reach shores across the ocean. Helped by the 1876 repeal of the restrictions on the issuing of passports, he worked on propaganda for the shipping companies worldwide, which had struck agreements with the Brazilian Government, either directly or on behalf of the colonization societies working in Brazil. Often the agent entrusted the recruiting to village notables, school teachers, municipal secretaries, parish priests and mayors. Those people, also led by the officers to believe that the migration offer was valid, encouraged their fellow-men to emigrate, promising advantages, property and optimal work conditions to families that in their home country were forced to farm land belonging to others, and to grapple day by day with debt, and make do by eating only polenta1. A perfect example of this is the poster circulated by the Genoa shipping agency of Clodomiro De Bernardis, the self-styled single agent in Italy entrusted with emigration to Brazil on behalf of Mister Caetano Pinto. Issued on February 4, 1877 in the Gazzetta di Treviso, the advertisement announced that:

> On arriving in Brazil, the settlers will be met by the agent interpreter and sent to their destination, where they will find a house, farm equipment, farm animals, seeds and all that they need, as well as board and lodging until the first harvest; the land will remain their property, the journey from Genoa to their destination will also be provided free of charge; however, they will incur a debt for the lodging, equipment and all that they need, but will have 7 years to pay it back. The Brazilian government, whose sole motive is none other than to attract honest,

> hard-working farm workers from whom one day it will have greater trade in its rich empire, since all can be gained from agriculture, does also this: if a family incurs a debt of 1,000 lire and pays it back within 3 years it will pay only half (the debt), in 5 years, two thirds, and in 7, the entire sum.
>
> The settlements to which we have agreed to send 100,000 settlers are: Santa Caterina, Porto Piuma (Spirito Santo), Rio Grande, San Paolo.
>
> As yet we cannot specify the time of departure, but in February things will be more certain, and probably there will be two crossings a month in postal steam-ships that will make the voyage every 23 days. If a family wishes to secure a place for itself, it is advised to send a deposit of 10 lire for each family.
>
> After the due calculations have been made, it is specified that every individual to arrive at the point of embarkation will incur an expense of about 24 lire if over the age of 12 years, and about half this sum if aged 4 to 12 years; and this sum is to be used for the journey from Conegliano to Genoa, and for the commission, board and lodging and embarkation costs.
>
> Nobody should leave home without being certain of his embarkation; nor should notice of his resignation be given to his master, and nor should he sell his belongings until he has received a letter of notification indicating the exact day on which he should be in Genoa; if he ignores the above instructions, he will bear full responsibility.[2]

The lines appearing in the newspaper failed to contain the clause, present in a circular sent later to their homes, stating that the migrants, on reaching their destination after being transported by sea free of charge, had to pay 15 lire a head for adults (children half this sum).[3] This was no small detail, since the average pay of a farmhand was 1 lira a day, but, notwithstanding this further payment, the offer must still have seemed heaven-sent to the poor Veneto peasants. So, on the crest of the wave of attractive promises, entire family groups departed in their search for a better and permanent settlement on the other side of the ocean, cutting all ties with their country of origin. Usually they had no land. If they had animals, equipment or furnishings, they sold them all in order to pay the agents' fees[4] and to cover the costs of their journey to Genoa, where, all together, they left from the city port – the elderly, men, women, and children.[5]

NOTES

1 The agents were committed to guaranteeing the owners from 400 to 500 emigrants per voyage, and to paying a fine of 40 to 50 lire for every missing emigrant (G. Erler, *op. cit.*, p. 62, note 8).

2 F. Meneghetti Casarin, *Treviso-Genova, andata e ritorno* [Treviso-Genoa, Return], The Benetton Foundation, Il Cardo, Venice, 1990, p. 18.

3 F. Meneghetti Casarin, *op. cit.*, note 44, p. 19.

4 Often the agents themselves took care of liquidating the miserable estates, the house and the patch of land of the emigrants, committing abuses and speculating in ways that can easily be imagined — generally beyond the pale of usury (G. Erler, *op. cit.*, p. 41).

5 A. Lazzarini, *op.cit.*, p. 318.

# 6
# From Promised Paradise to Real Hell

Awakening from the dream was sudden and brutal. The expected journey to Eden soon turned out to be exactly what it really was: Calvary. Tricked by the agents who made them pay far more than was due, the voyagers sometimes did not even find the embarkation they had been promised on reaching Genoa; they also fell prey to fraudsters of all kinds, from hoteliers to money lenders, from money exchangers to shop owners, all of whom were hell bent on grabbing their part of the illegal pickings to be made from the movements of the huge crowds waiting to leave. Often they suffered delays of weeks, sometimes months. Since, by law, the agent had to take care of the emigrant right up to his embarkation, he accompanied those in his charge to lodging houses that had agreed to pay him a percentage of their earnings. "One can only imagine the abuse committed under these circumstances. Many emigrants, made to come to the embarkation port days or weeks in advance, were put up in these lodging-houses, if that is what you can call them, where they were forced to spend their modest savings; once this money had gone, their sacks containing a few personal belongings were confiscated and the emigrants were thrown out onto the street. Sadly, all too well-known were the miserable scenes in the lodging houses and taverns in Genoa, in the streets and side-streets near Piazza Acqua Verde, Via Baldi, Porto Principe Amadeo or Via Doria."[1]

And then came the voyage, lasting twenty to thirty days, on steamships, sail-ships, or sail-cum steam ships during which, as recorded by Nicola Malnate, a police inspector at the port of Genoa:

> Robust families of peasants left Genoa healthy, but reached America as orphans, or as parents without children, corpses left behind in the sea.

Malnate also observed that the emigration law:

> …allowed the transportation of our emigrants to be provided in old carcasses traveling at 8 miles an hour, and the treatment on board continued to be hardly better than that meted out to slaves. . . with 2.25 (sq) meters of breathing space in the dormitories (the holds were reduced to corridors) for every adult or for two children under the age of 12, while hygiene would call for at least 25 cubic meters of air for every living being, adult or child…. In the summer, the steamship held closed within its flanks up to two thousand emigrants, without the benefit of navigation winds, lodged, piled up in disorderly undisciplined heaps, in a breathing space of less than two cubic meters of stale air for each of them. When one of these thousands has an infectious disease, the contagion soon spreads to the others.'[2]

The appalling conditions were also well described by Count Ferruccio Macola, editor of the *Gazzetta di Venezia,* who traveled to Brazil in 1893 on a steamboat called Washington "to get to the bottom of the more practical aspects of the matter of emigration":

> Now, what is the main cause of this slaughter, which fears neither statistics that expose it, nor newspapers that deplore it, nor philanthropists that bewail it … is it because the dead do not speak, while crass ignorance ensures the discretion of the living? It is the excessive crowding together that turns our national steamships for passenger transport into transporters of human flesh. Men, who have come to be considered goods, are diligently packed into the hold, until the last fraction of a cubic meter left over on board by tonnage is used up. That the goods transported in this way might seem somewhat damaged is of little importance; the main thing is to make sure that the cargo is on board, and that it can be demonstrated, whatever the case, that everything possible has been done to keep it healthy; this explains the perfunctory check by the doctor on board, who sails with the emigrants. And yet it would be so easy to

> find a remedy … let us suppose that the Government… in the interests of hygiene, ensured that the number of emigrants traveling on national steamships were reduced by a quarter, and that stricter measures were taken on board by, for example, making sure that the berths have a metal net at their base; if care were taken over the quality and quantity of provisions carried by the steamships; the medicines, etc.; the indecency, believe me, would soon disappear.[3]

Although it would have been easy to make improvements, nothing changed. Again, as observed by Macola:

> It is, in fact, odd when steamboats carrying so many hundreds of wretches manage to touch the first port for disembarkation with only one person dying on the way. One, two, or three dozen are counted dead, and swallowed up by the silent and indifferent sea; and it is children or the most exhausted, who are the victims of this sad blight. In only one voyage, Commander Zanelli told me, in spite of all the efforts made, twenty six children were lost! Twenty-six![4]

Until 1875, the flow of Italians to Brazil had been irregular. The year 1876 saw the beginning of emigration organized by the large shipping companies. This phenomenon, defined as "artificial" or "stimulated" emigration, was the fruit of a pre-paid journey stipulated in the contract together with other subsidies, rewards and enticements.[5] But if, as from 1875, the immigrant could make use of services, the Brazilian decree of October 28, 1885 was decisive, enabling the government to officially use a system by which contracts were drawn up with shipping companies to whom it reimbursed the costs of the emigrant's journey, and included free transport from the port of embarkation up to the chosen territory in Brazil as well as board and lodging for a specific period of time in the "hospedarias," which can be defined as temporary detention centers. Immigrants who arrived at their own expense were reimbursed the costs of the journey by the government.[6]

## Notes

[1] G. Erler, *op. cit.*, p. 41.

[2] M. Missori, "Le condizioni degli emigrati alla fine del XIX secolo in alcuni documenti delle autorità marittime"[The conditions of emigrants at the end of the nineteenth century in some naval authority documents], in *Affari sociali internazionali* [International social affairs], n. 3, 1973, p. 99.

[3] M. Missori, *op. cit.*, pp. 96–97.

[4] M. Missori, *op. cit.*, footnote 25, p. 101.

[5] In 1872 the first Brazilian census recorded a total of about 6,000 Italian residents. The effects of organized emigration were such that in 1876 alone "About 7,000 emigrants reached Brazil, and in 1880, the total was 50,000, after taking the repatriated and deceased into account. In 1887, 31,414 arrived, in 1888 97,730, in 1890, 16,230 and in 1891, 108,414, the largest number to be reached. From 1902 onwards, the number decreased to an average of 20,300 immigrants a year" (G. Erler, *op. cit.,* note 14, pp. 63 and 40).

[6] But only with decree no. 528 of June 28 1900 (the so-called Glycerio Law) were the real foundations laid for a true legislation for immigration and colonization, with the establishment of regulations designed to control the transportation and the introduction of immigrants into the territory of the Republic, and regulations governing the colonization of private property. In 1900, travel subsidies were abolished, thus paving the way for the transformation of "encouraged" or "artificial" immigration into "free" or "spontaneous" immigration. Some facilitations were to be re-instated in decree no. 6455 of April 18, 1907, which was designed to ensure selective immigration and, above all, to facilitate settlement on property (G. Erler, *op. cit,* p. 41, and pp. 44–45).

# 7
# Brazil At Last!

After the difficult, sometimes tragic, Atlantic crossing, further bitter disappointments awaited the emigrants on their arrival. Enticed by a journey free of charge and the thousand promises made to them, they reached Brazil to discover that the only true "benefit" had been, at a price — of risk and sacrifice — to have crossed the ocean free of charge. But there was no trace of anything else that had been promised them; no free equipment and utensils, provisions, houses, land, board and lodging in comfortable asylums or free journeys to the inland. Herded together for long quarantine periods in the hospedarias of Rio or San Paolo, exposed to violent epidemics like the one that had broken out in the latter city in 1888, the men were separated from the women and children — all of them were treated like animals.

The first thing the immigrants learned was that, after making a journey free of charge, they were bound to the contract that they had signed, more often than not without having been able to read it, obliged to go to the places chosen for them by governments or companies. Sorted brutally by the appointees of the fazenderos, who had no scruples about renewing the promise of heaven and earth or about directing the newcomers away from the places that they, invited by relatives or acquaintances, had originally intended to go to. Immigrants then had two possible destinies: become settlers and work their backs off by seizing tropical forestland in the hope of cultivating it, or work as farmhands on the coffee plantations.[1]

The conditions of these pitiful people were described well by Father Pietro Colbacchini, an apostolic missionary in Brazil:

> In not a few fazendas, the settlers, just like the slaves of old they have come to replace, cannot ever leave the owner's estate. They are not allowed visits from relatives

> or friends living in other fazendas, it is prohibited to go to church, also on vacations, in order to prevent them from coming into contact with their co-nationals, and even the postal correspondence is checked, and the secret betrayed.[2]

Confined and disillusioned, the immigrants had little choice: they had to stay put. They had their families with them, could not afford the return journey home for themselves and their family members, who were unwittingly the greatest obstacle to any plan of individual escape.[3] On the other hand, to prevent defections and repatriations, the owners carefully made sure that emigrants were unable to set aside any savings. The debt for the advance payment of the travel expenses from the coast to the plantation and the provision of food and equipment was rarely settled. Moreover, fines abounded for all sorts of trivialities and any acquisitions had to be made at the owner's store, at exorbitant prices, while the produce that the immigrant sold to the very same store fetched half its value. Payments of wages, constantly reduced due to the increased supply of labor, were spaced out and often deferred — and the frequent bankruptcy of enterprises themselves made the workers lose all hope of ever possessing the credit they had accumulated.

This was the fate of Veneto emigrants in Brazil. Preferred to other emigrants because they were tireless, peaceful, docile, submissive, moderate, neither demanding nor likely to form unions or organize resistance, let alone a rebellion,[4] the frontier atmosphere they encountered was quite different from the poor but civilized Veneto, and its setting was characterized by a savage economy and a natural environment that was treacherous and unknown, made even more dangerous by diseases such as yellow fever, the presence of ferocious animals and the resistance of indigenous populations against the settlements.[5]

Those destined for the colonial units, and who had dreamt of finding an estate and a home during their voyage, often had to start their new life in a wild land, covered by forest, without any streets or houses in it at all. They had to sleep stacked together in makeshift cabins until they had uprooted the forest and cleared some land to farm, and made space to build their first home. Under the imperial regime, the land belonged to the state and therefore settlement on the territory was achieved by the central government. In 1888, however, the government ceded 36 square *"leguas"* to each of the provinces that had been proclaimed states by the republic.

These unoccupied areas were destined exclusively for colonization.

Some units were founded by the central government, others by the single province with the consensus of the central government, yet others by private enterprises and by the immigrants themselves.[6] It was in the State of Rio Grande do Sul that the system of small holdings developed in a surprising way, the stretch of 36 leguas being split up into two large areas. One, more fertile, vast and flat, with less woodland, nearer to the State capital and irrigated by navigable rivers, was the so-called "Campanha," and extended to the "Serra Geral" foothills. The other, called "Encosta da Serra," lay on an immense, forest-covered mountain plain at a height of around 600 meters above sea level, intersected by narrow and deep valleys crossed by relentless torrents, thus consisting of an enormous stretch of so-called "devolutas" land, destined to the provinces by the central government for settlement but given back by the provinces that failed to deliver. Germans and Italians were the two main migrant groups populating the 36 *leguas*. Throughout the years, the two groups contended with one another for status as the main centers for the social and economic development of their respective colonial units, and for recognition as the principal commercial and industrial centers of the State.

So, while one area was occupied by German emigrants, its seat in the Rio Grande do Sul dating back to 1824, Italian settlements lay to the north of the German ones, in the "Encosta da Serra" region, which stretched for 7,817 square kilometers. The geographic position of the Italians' settlement was therefore far less favorable than that of the Germans.[7] Most of the Italians who reached Rio Grande do Sul were from Veneto.

Italian emigration true and proper to Rio Grande, started in the period between 1876 and 1877, and remained constant throughout the following years — at 3,000 to 4,000 — except for 1884, when fewer than 1,000 immigrants arrived. Following the measures taken by the republican regime, the number of arrivals continued to be above 7,000 in 1889, but dropped to 3,000 in 1890; in 1891 about 9,000 people arrived at Rio Grande do Sul and in 1892, about 7,000. Subsequently the number of immigrants went down to 1,000 a year, plateauing at a rate of about a hundred a year until 1914. From 1875 to 1889, Italians accounted for almost 90% of the total number of immigrants in this state. Three particularly important colonial units, founded in 1875, were destined to become not only the main immigrant-sorting centers of the region but also the cradle of social, economical, cultural and religious development for migrants in that period. These colonies arose on the river "Rio das Antas"; the first was Dona Isabel or Princesa Isabel (now Bento Gonçalves, the wine-producing capital of Brazil), the second, Conte d'Eu (now Garibaldi), and the third, Campos dos Burges, which in 1877 came to be called Caxias. Then came the colonies of Antonio Prado between the river Prata and the Pampa (Campos de encima de Serra) and that of Encantado between the Forqueta and Guaporé rivers. In 1892, the colony of Guaporé was founded between the Carreiro and Guaporé rivers.[8]

The colonies were divided into "linhas", each of which was divided into allotments. This subdivision was, however, made on maps without any regard to the topography of the area or the quality of its land. This meant that the formal equality in the size of portions was matched by a striking difference between one allotment and another; many of them being partly or totally unsuitable for farming. Each was, on average, 200 meters wide and 1,000 meters long, its shorter side usually being bordered by a river, a torrent, or a *"traversao"* or *"picade"* (i.e. a pathway). After razing the trees to the ground and burning them, the settlers managed to create areas for growing beans, corn and other cereals.

The great merit of the Italian emigrants was, however, that they introduced the cultivation of vines and corn[9] even though at first, in many Veneto settlements, corn was, as it had been in their home country, the staple diet, and was grown for up to ten years without a break on the same land as a guarantee against hunger.

The total lack of road connections delayed the marketing of agricultural products, which had to be carried to the bank of the river Caì to then be transferred to retail centers.[10] This enterprise took several days of traveling, transportation being provided by animals, which enormously increased the cost of products. This state of affairs led to the growth of another occupation of crucial importance, an activity that, as we will see later, was to play an important role in the story of the Dalla Costa family:[11] running the so-called "hay houses," post stations for resting and refreshing the draft animals, where, above all, transporters could stock up with enough hay to enable them to be self-sufficient while traveling long distances through the forest.

## Notes

1 A. Lazzarini, *op. cit*, p. 47.
2 A. Lazzarini, *op. cit*, p. 320–321.
3 However, there were episodes of rebellion and escape in response to the terrible conditions of life that the settlers were forced into. One emblematic case occurred in Paranà: "Our settlers stuck it out for a few months and then they rebelled. They had buried too many bodies; this was an accursed place. The owners and authorities did all they could to contain them by forcing them to respect the contract that they had signed, which allowed the Government to make use of them wherever it felt fit; but most of them would not budge. In the end the authorities let them leave. Some went back to Italy, emaciated, in conditions of extreme poverty. Most of them went on foot, through the forests, across the mountains into the inner region of Curityba. Nobody expected them: they had to shelter in makeshift dwellings, without anything; they had to beg for their own survival. Our people were very angry. They had been enticed with many promises, they had faced terrible challenges and yet they had neither a place to sleep, nor a patch of land to work. The situation was so tense that the authorities had to intervene, and they forced the settlers to surrender their weapons" (*La valigia dell'emigrante prev.. cit.,* pp. 36–37).
4 A. Lazzarini, *op. cit.*, p. 46.
5 "The first settler to be shot by an arrow was a certain Giovanni Baldassar from Tambre. While felling a tree in a cornfield, he suddenly lost his grip on his axe and slumped to the ground. This news traveled from cabin to cabin. The same year a woman who was alone at home in the attic with her three children suddenly heard two arrow heads hitting the walls. She hid her young daughter in the chest, and then, while moving extremely carefully, she gathered her two other smaller children into her arms and carried them to the nearest neighbours. When they felt safe, they went back together to the cabin. It had been raided. The child hidden in the chest had been slaughtered … There were other victims; and the soldiers were called in. War was waged between the settlers and the Indios (whom the whites called *'bugres'* and for whom the Veneto immigrants used the more simple name of *'bulgari'*), who felt threatened and tried to stop the intruders. Stealthily, they stole food, clothing, work tools and kitchen utensils, and they killed any individual settlers who happened to cross their path. On their part, settlers shot whenever they saw them (Indios). It was a systematic and cruel carnage, committed by gangs of professional hunters paid to do their dirty work by the Government and the colonizing companies. Settlers, too, used to tax themselves to pay for their [the hunters'] work. Any contact between the settlers and the Indios was always violent. The gangs of hunters entered

the forest and stayed there for a few weeks. When they found an Indios encampment they attacked it before dawn to take the Indios by surprise, while they were sleeping. Whoever escaped was chased and killed. Nobody was spared; except for the odd child that was taken amongst the whites, but usually died soon after. The victims' ears, kept in a sack by the leader of the expedition, were preserved in brine. On returning they were counted. For each right ear money was paid: it was payment for carnage. There were even crueler interventions. One photograph, which can still be seen in some families, shows a hunter on horseback. On the saddle, left and right, hang bunches of heads of Indios killed and carried with pride, just like war trophies" (*La valigia dell'emigrante, prev. cit.,* p. 35).

[6] G. Erler, *op. cit.*, pp. 45–46.

[7] *Ibidem.*

[8] From 1900 onwards, due to the huge increase in the population, internal migration took place on a massive scale. The settlers, but above all their children, left the primitive settlements to go plow new land. From half way through the twenties, and above all the thirties onwards, new colonial units were created by the immigrants' offspring (G. Erler, *op. cit.*, p. 48). Regarding the colonization of the Rio Grande do Sul by the Veneto people, see also Ulderico Bernardi, *A catàr fortuna. Storie venete d'Australia e del Brasile* [Searching for fortune: Veneto stories of Australia and Brazil], Neri Pozza, Vicenza, 1994.

[9] In time, the region of Rio Grande do Sul became an important area for the production of fine wines that were in demand worldwide.

[10] G. Erler, *op. cit.*, p. 51.

[11] Part of the family in fact, as we will see later, emigrated to Rio Grande do Sul, settling in Guaporé.

# 8
# Fire, France, America

Of course, in the late summer of 1886, Giovanni Dalla Costa could never have had any inkling of the harsh reality and difficult conditions awaiting Veneto peasants on the other side of the Atlantic, although he already was none too approving of emigration to Brazil. He would never have crossed the ocean just to keep on working as a peasant, as he was doing in Costa. Rather, he would go to America in search of gold; to find it, all that people needed to do was scratch the ground a bit and, with a touch of luck, they came home rich, seriously rich — much better than staying in Italy and breaking your back trying to scrape up a living.

But Giovanni was young, and these were only youthful, light-hearted dreams shared with friends, nothing more than the whims of a healthy and strong eighteen-year-old who, like all other youths of all other generations, let his mind roam towards horizons far greater and mythical than those existing in the real everyday world. Certainly he could never have had the remotest idea of the misadventures, and adventures, that destiny held in store for him and his family. Nor could he have known how prophetic his words would soon prove to be.

One night that fall, the Dalla Costa family home was devastated by a terrible fire. This event was doubly disastrous, since it happened during late harvest time, when the attic, the cellar, the hay-loft and stables were full of "God's bounty," the food needed for the family to survive another winter. Everything was destroyed and the family was plunged into desperate circumstances that called for extreme decisions. In other times, this misfortune might have been considered a plain nuisance, but in the particular setting of those times in the Veneto, it was a tragedy. What could be done? Money was needed, and it was needed now, but Francesco was doing his military service, and Giacomo was only sixteen.

There was no alternative: it was up to him, Giovanni, to somehow find work and do his utmost to set matters right. No sooner said than done, and somehow he managed to scrape twenty lire together and left early in 1887 for France where, a short distance across the border, he started working whenever he got the chance — mainly in coal mines, a bit here and a bit there, wherever laborers were in demand. Conditions were hard and risky but this was the only way an Italian emigrant without means could earn as much as possible in the shortest possible time.

During the time spent up and down the mines, Giovanni came to learn the basics of mining techniques, skills that would serve him extremely well in the years to come. He did casual work and, sturdy and willing as he was, managed not only to make ends meet, but also to send something back home to his family. Little more than a year was spent in this way, when the news from home left little room for hope. He found out that his sister Maria Anna had given birth to her third child, Maria Regina. He also learned that he would not be drafted for military service straightaway — he had been registered in the third category and was provisionally exempt because his brother Francesco was already serving in the army. But, above all, he realized that his efforts were not enough to resolve the situation and that, day by day, things back home were going from bad to worse.

Disappointed and bitter, towards the end of summer 1888 he took the decision that had long been brewing in his mind: leave for North America to look for gold, make a fortune, and really help his family members, who were in dire straits. So there Giovanni was, completely alone, at the port of Le Havre, stepping onto the ramp of the transatlantic ship leaving for New York, ready to be a protagonist in a story that had started two decades before he was born: the gold rush.

## Notes

[1] Recollections of Francesco Dalla Costa, first son of Giovanni, recorded by his wife, Maria Ghidelli.

[2] Parish archive of Pederobba.

# 9
# The Gold Rush

In 1848,[1] a Swiss, John Augustus Sutter, was living on the shores of the Pacific, in north San Francisco, more or less where the city of Sacramento now stands. On fleeing from debts incurred in his mother-country in 1839, he reached California, which, in those times, belonged to Mexico, like Oregon, Texas, South Arizona and New Mexico; he soon obtained Mexican citizenship.

In the same year, at the end of the war between the United States and Mexico, under the Guadalupe-Hidalgo peace agreement, Mexico had to give the United States all her territories north of Rio Grande, right up to the border with Canada. The expansion of the north American nation which, up to then, had never exceeded the confines of the Rocky Mountains, made a huge leap forward by suddenly acquiring an enormous area, with few indigenous inhabitants; this virgin world was there to be explored and colonized.

The population in the United States had grown from just over ten million in 1820 to almost thirty million by 1848, thanks to the huge migratory flow from Europe. The need for new land was now getting more pressing; on top of this, the survivors of the war with Mexico needed to be provided with work. So, what better occasion for the American governors than this new, immense territory to satisfy the hopes and the vitality of a mass of individuals in search of a future? But, to persuade the greatest possible number of people to leave for the unknown West, a new reason had to be found, a unique opportunity, an irresistible incentive.

Chance had it that exactly in that moment Sutter, on deviating a flow of water toward the timber mill that he had built with his partner, James Marshall, saw something shiny in a crack in the granite; he took it out of the water and, after clamping and hammering it, came to the conclusion that it was gold. Although he made every effort to keep this news secret, on March 15, the *San Francisco Californian* came out with a short piece on it. This was enough to ensure that half-way through the following June three quarters of all the male citizens of San Francisco were on their way, in search of gold. The publication of newspapers was halted because of the staff shortage and even the meetings of the municipal council were cancelled. Bit by bit, the glad tidings trickled across the state borders and, by late summer, hopeful prospectors had rushed away from Hawaii, Oregon, Mexico, Chile, Australia and China to reach the spot.

On the east coast of the United States, the official go-ahead for the gold rush was given by President Polk who, on December 5, 1848, during his Annual Address to Congress, declared that the mines, whose existence was already known about when Mexico sold California to the United States, were more vast and precious than could ever have been imagined. He included a report received from the military governor of the lands, in which it was written that the gold-bearing riches had just been scraped from the surface, and that in a single day a gold miner could earn double the monthly wages of a soldier; moreover, he said, the gold in California would have repaid more than a hundred-fold the cost of the war with Mexico.

The dice had been cast. Thousands of people of all social classes, from north to south, decided to accept the challenge, leave their home and their job, if they had one, and organize their departure as fast as they could. Not only this, already in January 1849 ships were leaving all the European ports for California; the same was happening in the Asiatic ports, particularly the Chinese ones.

Old disused vessels were smartened up as much as possible to meet the enormous demand — with the inevitable corollary of accidents and shipwrecks that still failed to discourage these departures. And if, for those who crossed the Pacific, the journey to California was long and dangerous, but took them straight to their destination, obstacles of a completely different nature awaited those who left from the East coast of the United States or, even worse, from the ports of old Europe.

For those who had to reach California from the east, there were only three possibilities: either travel by land following one of the different routes that crossed the central and southern parts of the United States or the northern areas of Mexico, or across the sea, circumnavigating Latin America, otherwise reaching Panama by crossing the isthmus overland, resuming the sea voyage towards the north along the Pacific coast.

It was like choosing between the devil and the dark blue sea.

## Note

[1] The information contained in the following pages regarding the gold rush are taken mainly from the work in two volumes of W. Greever, *La corsa all'oro* [The gold rush], Il Saggiatore, Milan, 1966.

# 10
# We're Leaving, Whatever

Overland journeys were made using wagons pulled by oxen, mules and horses. Usually a caravan was organized by various groups of three or four people after pooling their resources to buy the necessary means and supplies, but also simply by family units or by newlyweds.

May was the most popular time of year for leaving because in the months that followed, the grass along the way would be long enough for the animals to graze. Thousands of kilometers of journey awaited these courageous people, steep climbs uphill, downhill descents made even more dangerous by the poor braking systems of wagons, which often hurtled along so fast that the wheels broke, or turned over with disastrous consequences. The direction to keep was also a constant uncertainty. Often there were deviations from the right path, discovered only days later, a hitch that usually meant going back on one's tracks for several miles.

Passing across or along rivers was another risk, no matter how skilled the coachmen were. A further constantly lurking danger was attacks from Indians. These were not yet the times of war true and proper, but skirmishes were frequent, often due to theft and broken agreements. In the years spanning the mid-nineteenth century, another threat was cholera, which struck more readily where many caravans had settled together simultaneously, making the water wells infectious.

So this was a hard and hazardous journey towards hope, during which the diet consisted of smoked lard, rusks, a lot of coffee, and sometimes *focacce* [flat cakes] baked over makeshift night-fires. A popular alternative was buffalo meat, which was sometimes cooked and eaten on the spot or, more often, cut into strips, smoked and kept for emergencies. Yet, buffalo hunting was neither simple nor risk-free. Often the hunters were thrown, and sometimes wounded seriously while the horses, free at last, took advantage of the situation, disappearing over the horizon together with the buffaloes.

So the travelers crossed vast plains, marshy stretches and areas with quick sand, thick woods and forests, endless deserts and rough mountain chains, on a journey that was meant to change their lives, but more often than not ended up changing the travelers themselves. Any tiny delay meant that they were suddenly in the thick of winter, and on foot, their wagons and animals being lost in the midst of snow storms on the high mountains. On the other hand, if they failed to take the due precautions when pressing along the Spanish tract opening out from Santa Fè into southern California, travelers had to endure the nightmarish sun and thirst in Death Valley. It was no accident that the people who reached California were described as "exhausted individuals put through unimaginable hardship."[1]

As stated, the alternative was to travel by sea. And this decision was very commonly taken: in the first stages of the gold rush, ships with the American flag had almost completely disappeared from the ports of the rest of the world. They were all busy shipping to California, packed with men and women who had a dream firmly fixed in their minds: making a fortune.

But these "sea" opportunities, as well, were anything but painless. The voyage from New York, Boston, Philadelphia or Baltimore to San Francisco lasted over six months. Half a year in vessels that more often than not were cargo or whaling ships hastily adapted for the transport of passengers, six months during which they passed from the temperate climate of North America to the unbearable heat of the Equator and the Tropics to then descend into the frozen confines of the South Pole, overcoming the constant danger of being shipwrecked, and the tempestuous waters of Cape Horn, to then go up again toward the tropical heat to at last reach the Bay of San Francisco.

Another option was open to those who wished to avoid such a long sea passage: go overland across the Isthmus of Panama to re-board on the Pacific coast. But there were ifs and buts in this case, too. Above all, the difficulty of the crossing, which called for less than a week but required a few days aboard bungos, indigenous canoes maneuvered by local boatmen along the Chagres river, and then further days of tough journeying along the long pathways that were extremely narrow and tortuous, often hardly viable –scorching hot in the dry season, they turned into fast moving mud during the rainy season. These adversities, however, were not the only problem. On reaching Panama, the gold seekers had to wait up to four months before finding a place on a ship to take them North. This hindrance was to decrease with time, but it persisted even after the inauguration of the Panama railroad that, in 1855, took migrants from one ocean to the other in less than four hours. But the figures speak for themselves: in 1849 15,597 gold seekers arrived in San Francisco by sea via Cape Horn and 6,489 via Panama. In the following years, and even more so after the railroad had been set up, the number of Panamese grew until it plateaued at around 15,000 to 20,000 a year.

## NOTE

1 W. Greever, *op. cit.*, vol. I, p. 42.

# 11
# New Cities Are Born

Each year there were therefore thousands of new arrivals in the land of gold. The first stop was an agglomerate of tents and shacks that in 1849 alone saw its population soar from about 3,000 to 20,000 souls. This city, San Francisco, had various quarters, including Little Chile for the Hispanics, Little Germany, Little France, Little Chinatown and then Sydney Town, inhabited by English ex-convicts who had been deported to Australia.

There was enormous speculation, a place to sleep was expensive, and the price of a shed was at least five times higher than that asked for on the Atlantic coast, but, on the other hand, nothing was missing - there were even Chinese restaurants with their distinctive yellow triangular flag offering good prices.

In those times, San Francisco was a city that grew out of nothing. Overlooked then, as it is now, by Telegraph Hill, it was marked by frenetic activity. The equipment and animals required in the search for gold were bought and sold, hills were flattened, roads were built even if in wintertime they would turn into rivers of mud, walkways were set down using all types of materials, and disastrous fires were put out although without managing to prevent hundreds of tents and shacks from catching fire.

Above all, business was done, and this enabled the city to grow - and to slowly but surely take on the appearance of an urban center, true and proper, something quite different from the temporary stopping place for occasional groups in search of gold that it had been until recently. The effects of the gold-rush started to show, not only in San Francisco, but throughout California. In different parts of the state, new strategically located cities were emerging fast near the mining areas, the most important being Sacramento, and then Marysville, Downieville, Nevada City, Grass Valley, Coloma, Placerville, Jackson, Mokelumne Hill, Sonora, and so forth. An incredibly vast territory, until then almost totally empty and wild, was rapidly evolving and taking on a shape of its own.

Within a few years, a solid network of transport between the gold fields and the various cities and towns was activated. River boats with side-wheels ran up and down the Sacramento River — from San Francisco to Sacramento and back. Companies owning thousands of mules organized convoys of about 50 animals, each with a load of over 150 kilograms, directed towards the various mining areas with the utensils and tools of the prospectors. Then, when the paths were transformed into roads, came the moment for wagons to carry goods and stage-coaches to move people. Soon the services became regular and frequent. At the end of 1851, for example, every day six mail-coaches left Sacramento for Marysville, two for Coloma and one each for Nevada City, Placerville, Auburn, Stockton, Drytown and Jackson.

A few years later, in 1856, the network of standard routes covered a distance of more than 6,000 kilometers. Then, on April 3, 1860, the first Pony Express left the terminal railroad station of Saint Joseph in Missouri to reach San Francisco, covering the impressive distance of about 4,000 kilometers in only ten days. This phenomenon was legendary but short-lived: only a year and a half later the first transcontinental telegraphic line was inaugurated, thus putting an end to the transportation of letters on horseback.

## 12
## The Search

The real protagonists in this rapidly changing universe were the gold diggers themselves, for they were committed to finding the life-changing vein of gold. Although all and sundry had a rough idea of where the riches were, the true collective bet was on the gold deposit worth exploiting, and the bet was fundamentally based on chance. As the gold-diggers would say: gold is where you find it. One story that went round was about how some diggers were at a friend's funeral and, bored by the minister's sermon, they started fiddling about with the earth dug up for the grave. Seeing their excitement, the minister took a clod of the earth in his hand and, on realizing what was in it, he started shouting his head off, exclaiming that such a rich deposit had never been seen before. Another tale was that of a digger who had invested all he had in searching a particular area only to find nothing. Out of sheer desperation, he killed his wife, his three children and himself. A few days later, on that very same spot, just a few feet further down, other diggers found a deposit worth millions of dollars.

Whether lucky or unlucky, these men were constantly on the move in their search for the richest vein of gold; like grasshoppers, they would cover and explore those lands, getting to know their each and every fold, putting up shacks and houses, towns and cities, bringing their presence and the signs of their work into every single nook and cranny.

Most of them were experienced miners, from Cornwall, Wales, Germany, Spain, France, Italy, Chile, Mexico, from the coal and iron ore mines of Pennsylvania or the lead mines of northern Mississippi. Men who knew the skills required for extraction, even though, in the early days, not that much experience was needed, since gold was there for the taking. A rule of the thumb was to follow the riverbeds, whether they were in full flow or had dried up. But, of course, this approach could not be used in every single case: in ancient geological eras rivers might have run completely different courses.

In 1848, many different techniques were used to search for gold, the choice depending on the richness of the deposit and on the availability of cash to buy equipment. The first, most rudimentary way - the only one that did not require water - was to extract the gold from the cracks on the banks of either a running or a dry stream. This operation was done under direct vision, using a knife or any other sharp instrument. The second option was the bowl-system. One person did this task using a flat-based container, a sort of concave tray or sieve, which he partly filled with soil; then water was poured into the sieve, and it was swirled around; the tiny whirlpool created by rotating and tilting movements sifted the earth and sand out. Because of its higher specific gravity, any gold dust settled on the bottom of the shallow recipient. Each procedure would take about ten minutes. One person would repeat this about fifty times a day, but it was only worth the effort if deposits were plentiful.

A more efficient way, for which two or three people were required, made use of the "rocker", a gadget that was like a baby's cradle, basically consisting of a chest, roughly one meter long, half a meter wide and ten centimeters high, with two supports, a handle on the side and, on its upper part, a hopper with a grid. The soil was placed on top of it and the gadget was rocked with one hand, generous quantities of water being poured onto the material with the other hand, thus allowing a first separation from the larger stones. Beneath the grid was a slide made of canvas or wood, designed to capture the tiniest fractions, while on the bottom of the rocker were another two or three wooden inserts which further sieved the slush before it finally came out of the hole at the base. Once the procedure was finished, gold dust would be recovered from the slide, the inserts and the base.

A further development that, however, required a team of three to six people, was the so-called Long Tom, a non-rocking device that consisted of an upper hopper on which the soil was placed, and a board, three to ten meters long, with closed edges and several transverse bars. For this system to work, a constant and strong flow of water was required; this was usually supplied by a hose that siphoned water from a basin located at a higher level. Following the first selection, made using a metal sheet with large holes, the material would slide along the board together with the water while, bit by bit, the gold powder would settle on the bars.

The sluice was similar to the Long Tom, but had another set of planks and bars at its end, and could filter the material further, thus extracting more gold. The number of men working at the sluice varied, ranging from a minimum of five to a maximum of twenty. Additional improvements were made by using mercury grids, which captured gold for amalgam. Yet no matter how many improvements were made to these extraction techniques, all the systems were home-made and somewhat slip-shod. Deposits were getting harder and harder to find with low-cost techniques. In the early fifties, in fact, mining companies with huge capital and industrial extracting systems had decreed the demise of the first wave of do-it-yourself gold seekers.

The new industry, which did things on a grand scale, re-routed rivers by creating canals, and it dug into dried river beds. The novel approach was as efficient as it was devastating: in 1855, about 10,000 kilometers of artificial canals had already been opened up, and promising waterways, like the American River, were forced to flow for thousands of kilometers along lines that had nothing to do with their own natural course.

Another industrial system was to dig deep, or to dig long tunnels by the hillsides in which gold-yielding gravel had been found, the material obtained being washed in the nearest course of water. Then, still in the search for the greatest possible bounty with the least possible effort, men started washing material in canals that had been dug out by the side of a hill. Shovelfuls of sand and gravel were thrown into the continuously flowing water, the material being filtered from the natural obstacles present in the artificial canals. And finally, hydraulic excavation of the hills, true and proper, was started, even though it meant destroying the underlying harvest, uprooting orchards, damaging the foundations of houses, and sometimes, even forcing entire villages to leave. What did this involve? Literally razing to the ground a hill considered economically viable for obtaining gold by sweeping it away into a valley, with the help of explosives. The material, carried along with the water in washing canals that were several kilometers long and arranged in the valley bed, gradually deposited its gold content. These systems, extremely damaging to the environment, were profitable. A hundred-fold more cost effective than the old rocker, this new approach soon made the individual gold digger obsolete: he was transformed into an employee of the mining companies.

The most evident upshot of this sudden transformation in the productive organization was that in California, already in 1858, many gold seekers no longer had a job, and were looking for work. Work that they found increasingly often in the gold quartz mines, where the extraction of gold was achieved using the new modalities, digging activities being conducted on the land surface or in deep pits using pick axes, hand drills, bludgeons and small quantities of explosives; the gold was then extracted from the rocks using costly crushers, grinders and chemical procedures, based on mercury and chlorine. Thus, in just ten years, many of the men who had taken part in the first Californian gold rush were reduced from being free prospectors to paid miners. But this was the other side of the coin, less palatable and therefore less easily acknowledged by those who wanted to try their luck. Fortune hunters continued to identify themselves with the man who, in November 1854, at Calaveras found the largest ever gold nugget, weighing 70 kilos, worth more than 30,000 dollars.

## NOTE

[1] W. Greever, *op. cit.*, vol. I, p. 95.

# 13
# Gold Gives Birth to a Nation

This was life in California in those years, with its ups and downs, fortune and misfortune, the hard work of everyday life and the myth of the Bonanza, the large gold deposit that changed the destiny of the man who found it. Eighty per cent of the population in California in those times was made up of young men, aged 18 to 35. The boldest, strongest and most adventurous men from all parts of the world had come here to make their fortune, and they were not about to give up, even in the face of heavy adversity.

The diet was incredibly monotonous. Bread, salted pork, dried beef, occasionally beans, rice, game, and every now and then, potatoes, onions, apples and peaches. In the initial years, from 1846 to 1856, the diseases going round were sometimes hard to recover from, and contemporary journalists reported that about a third of those who reached California ended up catching cholera, typhus, malaria (contracted in Panama), meningitis or rheumatic fever. The sand storms caused coughs and infections of the eyes and throat, and lung complaints were widespread.

The lodgings reflected the type of life led by these men, who were deprived of female company (there was one woman to every eight or ten men in the more inhabited areas); above all, the seekers had no hope of staying long in any particular place since the search for gold constantly pushed them onwards. So they lived in wagons, tents, caves and shacks, and slept on rough piles of straw.

Also alcoholism was common — so much so that in 1853, in San Francisco alone, there were 537 liquor stores. There were also professional gamblers, ready to arrange betting in the local mirror-lined saloons that they had set up to attract the prospectors after they had spent months on end in isolation, with large paintings of naked women, and a real flesh and blood woman every now and then.

So, was California hell? In some ways it was, but this world was bursting with enthusiasm and fresh energy, its natural environment was still unspoiled and full of God's bounty. Here the frontier spirit reigned, the social hierarchy had yet to be formed and the basic respect paid to individuals depended on the fact that anyone, however poor, might suddenly come across the right vein, the deposit worth tens of thousands of dollars, and become stinking rich. The gold rush was the back bone of the original Californian development, but it innervated other lines and ganglions of an economic system that was varied and ramified; it was the ideal springing board for a process of growth that is still going on now, after more than a hundred and fifty years.

Not all the seekers, in fact, continued to search for fortune by digging and sieving. Many preferred to devote themselves to farming along the valley of Sacramento to the North and in the Valley of San Joaquin in the South. They grew barley, corn, rye, fruit, including grapes, and, above all, bred cattle, horses and sheep. Others were involved in trade. Apart from the larger towns and the cities in which stores of all kinds were opened, every mining field needed at least one store for equipment, food and clothing.

Transport was booming, and companies working in this sector were increasingly well equipped, guaranteeing a steady flow of goods and people. Saloons and hotels also sprang up in ever greater numbers to provide refreshment and to receive the numerous travelers, who continued to flow in from abroad or who moved inland in the search for new opportunities. Also the number of physicians, not only quacks trumpeting out universal remedies, grew in number, alongside the hospitals, which were often run by brotherhoods and religious associations. New publications abounded: in San Francisco alone, in 1853, nine morning newspapers and three evening ones were issued, three weeklies and two fortnightlies. Shows were organized, and there were concerts (in 1851 the first opera company came from Italy) and dancing evenings, even if few women attended. The plays shown ranged from Shakespeare's Macbeth to Boccaccio's farces. By spending a few cents at the theater, lone men could enjoy themselves and laugh their heads off with heavy doses of risqué quips and dirty talk. And the management of public order and law was evolving – once the seekers had to do things off their own bats, but now courts were being founded and there were vigilantes and rangers who, using somewhat questionable methods, enforced respect for the rules that the different communities had imposed at a local level with codes of conduct that were as clear and concise as they were strict.

So a new world was in the making and, starting with the unspoiled wild land of California, in the space of a few years the power of gold managed to forge a well structured economic and social reality ready to be admitted into the heart of the Union as a State in which slavery had been abolished.

# 14
# Beyond California

After about ten years of great adventures, the rush for gold was slowing down in California. Now solidly in the hands of the large companies, the gold mining industry continued with its methodical daily work, but it had lost its original legendary character to become only a part, albeit a substantial one, of a vast, well organized productive universe, which was growing in all its sectors, and offered plenty of chances to earn money but no longer held the golden opportunity of making a life-changing find.

So where could the hunger for gold and fortune, which continued to arrive in these parts from the east of the United States, Europe and the Far East turn to? The hunger was no longer directed toward California, but had left, traveling across Placerville following the original immigrants' route, and had now alighted in Nevada, in the sierras close to the city of Reno, in an emerging town of tents and cabins christened Virginia City by an assembly of its first inhabitants. In 1859, the mine of Ophir had been discovered here; rich in gold minerals, its yield was, however, always mixed with an annoying blue substance. When, about a month later, this material was at last analyzed, it was found to be silver, of great value.

In 1860, as soon as the first spring warmth had started to melt the snow on the passes, veteran seekers from the Californian countryside came to the area along with others who had just arrived from all over, first on foot carrying their baggage on donkeys, then on horse-back and then in streetcars — all of them were scattered around again like ants, full of energy and ready to picket the rocks. In the course of that single year about four thousand concessions were granted within a radius of about 60 kilometers from Virginia City.

The fever had returned, this time for gold and silver. Experts in reckoning the value of a gold deposit, the seekers soon learned how to identify silver, too. The quartz had to be pulverized, the lighter mineral eliminated and the rest put into a recipient together with nitric acid; all the contents were then boiled until vapors could be seen — first red and then white. Then the compound was left to cool, put into a test tube and salt was added; if the liquid immediately became milky, it contained silver. If there was still a shadow of a doubt, it was enough to add some hydrochloric acid, hold the test tube to the sunlight for a few seconds and see if the contents turned purple. This was incontrovertible proof of the presence of silver.

Here in Nevada the shift toward the development of a more organized social and urban framework was far more rapid than in San Francisco and the first cities of California. To have an idea of the proportions of the phenomenon suffice it to say that in October 1860, a few months after the great inflow, in Virginia City there were already four tobacconists, three pharmacies, two stationary stores, two fruit stores, a hotel, four butchers, nine ovens, seven forges, three tinsmiths, a saddler's store, a gun smith, seven hosiery stores, two carpenters, a whitewasher, a tailor, three clockmakers, two barbers, six doctors, a dentist, eight lawyers, two post offices, a surveyor, five mediators, an auctioning room, a tailor for women, four sewing shops, ten stables, twenty-five saloons, nine restaurants, seven lodging houses, ten launderers, a theater, a night-club and five timber merchants. More living areas were built in the surroundings, and new roads were laid down across the mountains toward California. This tumultuous, paradoxical period was marked by assaults on stage coaches and the institution of new courthouses, and by the first experiment with dynamite, which was to replace gunpowder; but there were hundreds of victims, too, due not only to mining accidents but also to the fact that the drinking water contained too much arsenic, graphite, iron sulfate, and gold and silver salts. Strongly accelerated by gold and silver, the process of development in the area surrounding Virginia City gave body in a short time to a new state, Nevada, born in September 1864 as the tangible result of the mining activities.

The deposit, because it was really a single enormous one, as the miners were to find out in time, was truly large and rich. Organized and run by the mining companies, the searches in the shafts — extremely deep round holes — were made by miners who went down in rudimentary lifts to a depth of up to three hundred meters, where the rock was so hot that it could not be touched by the naked hand, and they achieved their goal a full sixteen times over, fetching sacks of minerals that went down in history by the name of Bonanza. The most legendary of these finds was the Big Bonanza, discovered in 1873 a little over 100 meters deep, worth the beauty of 150 million dollars to the company that had it on concession. Other discoveries were less dramatic and more hard to come by, but hopes were so high that the digging into Mount Davidson and the lower peaks of the Washoe mountain chain continued for over twenty years. In January 1881, the material extracted in the area had yielded 45% in gold and 55% in silver, an overall wealth that, as had happened in California, was to trigger the growth of a solid and well organized economic system.

The gold rush raged on in a similar fashion in other unexplored territories, like wild fire across the prairies, constantly fueled by the wind of hope of a big find. In 1859 it reached the tracks that led to Colorado, and arrived in the surroundings of Denver, at the feet of the Rocky Mountains. The seekers also opened the way in these areas, alone or in small groups, spurred on by the myriad stories, most of which were frontier legends and left large margins of doubt. Some of these men could afford it, but most had sold everything they had just to be able to leave. Some traveled in well equipped coaches while others went on horseback, with a tiny cart or a wheel-barrow — otherwise they went on foot with a sack on their shoulder.

The roads from Missouri were arduous and dangerous and nor was Denver organized to cope with the stream of arrivals. Of the 100,000 or so prospectors that left heading west from the central United States in the spring of 1859, around half reached Denver and, of these, half went straight back to where they had come from. But the search continued, accompanied by huge disappointment and small comfort, but enough success to prompt a repeat arrival en masse of a wave of young seekers in 1860. This enormous movement of populations was recorded in the census made in 1861 in the area, that counted more than 25,000 inhabitants three thousand of whom were in Denver, the others being spread out on the mountains and their surrounding areas, digging, exploring, opening up pathways, putting up shacks, and sometimes finding gold and silver. Notwithstanding the enormous difficulties and the poor hygienic conditions (in the late seventies the likelihood of a newborn baby surviving to the age of three was 50%), the process of development induced by gold was thriving here, too, and, in 1876, Colorado was officially proclaimed a State.

Onward, onward and onward again, like an unstoppable river, the front lines of the gold army marched on to Montana where the search for the precious metal was intertwined with the events of three wars against the Indian tribes of the Sioux, the Nez Percé and the Crows before ending, in 1889, with the birth of a new state, Montana. The search continued in the Black Hills, south west of South Dakota, a hilly area considered sacred by the Sioux, who would collect grass, fruit and berries there, bathe in the therapeutic spring waters, cut timber to build their homes, and extract stones to transform into deadly arrowheads. In 1868, after a war had broken out because of a new pathway being built towards the mines of Montana that crossed the plains disturbing the hunting by the natives, the United States and the Sioux signed a treaty recognizing a territory that included the Black Hills as an Indian reservation, which no white man could enter without a special permit.

But, on realizing that the Sioux bought their provisions with gold nuggets, gold fever also arrived here, followed hard on its heels by the troops of General Custer who, entrusted with controlling the situation and inspecting the area, noticed not only the presence not only of wonderful pasture land, but also of gold. The first news of the discovery, published in August 1874 in the Chicago Inter-Ocean, paved the way for a long-lasting conflict between the seekers who tried to enter the territory, the soldiers who were obliged to hold them back and the angry Indians, increasingly frustrated by having to stay within the confines of the reservation while they had to witness the growing profanation of their sacred hills. The struggle continued for a couple of years, until spring 1876, with the start of the military campaign which, notwithstanding the heavy defeat inflicted by Sitting Bull on General Custer at Little Big Horn, eventually led to the defeat of the Indians, who were forced to accept an agreement that obliged them to leave the Black Hills; they retreated to a reservation along the Missouri river, in Oklahoma.

The gold rush in the Black Hills was the most adventurous and famous of all, but probably the least propitious. No great Bonanza was found here. But in places where before there had been nothing but an Indian paradise, various centers and cities were born, including the greatest, Deadwood, where legendary characters like Bill Hickok, the great gun-man, and Calamity Jane, the legendary outlaw, used to roam. An important agricultural and animal farming activity was also set up and, in the late nineteenth century, more than 300,000 animals were to be seen grazing on the Black Hills, but there were no longer any Indians in sight.

# 15
# Giovanni Reaches America

Enterprising and willing, Giovanni worked as a cabin boy while on ship — crossing the ocean to reach this "new" America. This humble job enabled him to earn enough to recuperate a large part of the money paid for his ticket, and to reach the United States with a little money in his pocket. On arriving in New York, he probably got onto the transcontinental railroad that took him to San Francisco. But here he did not find what he was looking for.

This was toward the end of the year 1888 and the cluster of tents of gold seekers, where only forty years before he would have probably settled for good, was no longer there. In its stead, a large modern city had sprung up, and Giovanni soon realized that here the opportunities for a poor young emigrant were few and far between. His dream was not to find a job anywhere in order to simply survive, but to have a grand adventure that would make a rich man of him. The first few pieces of information that he gathered were enough to make him realize that really he could do nothing more than continue with the only thing that he knew how to do, other than working as an agricultural farmer: mining. To do this, however, he had to leave San Francisco and head up North, first towards northern California, where he worked briefly as a miner, and then on again, towards the state of Washington, on the Canadian border where, he had heard, there was gold, silver, lead, zinc and coal mining, with a consequent need for miners.

In 1889, Giovanni Dalla Costa reached the mining area south of Seattle where, once again, he was confronted with a situation that was not what he had expected at all, with individuals, alone or with a few companions, digging the ground in their search for a fortune. In the state of Washington the first deposits of gold dust had been discovered in 1882. In 1885 there were silver, lead and zinc mines that, in 1886, attracted large investments from mining companies. When Giovanni Dalla Costa arrived, the only way of making money was therefore to become a company employee. This prospect held little excitement for him, calling simply for hard work, day by day.

This is where Giovanni probably was when, in early 1890, he learned from his parents, Luigi and Teresa, that his family's situation had precipitated, and that the sums of money that he sent back home were simply not enough to allow them all to continue to live in Costa. His brother Francesco was back there, but this was not enough. Teresa had been to Montebelluna to ask the bank for a loan, but all in vain; since financial help was not forthcoming, the only option still open to the family was to emigrate.

Thus, only two years after circumstances had forced Giovanni to leave Pederobba, his parents and brothers were touched by the same destiny. It might have been otherwise had they been spared the disastrous fire, but now they decided follow in the sad footsteps of thousands of other peasants who had traveled toward the unknown, and emigrated to America. Giovanni's parents sold everything they had, including animals, work equipment, pots and pans, crockery and also a few pieces of jewelry, so as to collect enough money for the journey to Uruguay, the only possible country because Luigi was too old to qualify for the emigration program of the Brazilian government, which provided the journey free of charge.

The parents left together with their younger daughter, Elisabetta, and youngest son, Gaspare. On March 16, at Genoa they boarded the vessel of the Rubattino shipping company bound for Montevideo. It is believed that they had no clue as to what would befall them on their arrival in South America, nor could they have known where they would live out the rest of their lives. Francesco and Giacomo, on the other hand, stayed in Costa and tried to scrape up enough money to follow Giovanni in his path towards the countries across the borders, Austria, Switzerland, probably France.

So the Dalla Costa family, which, according to reliable sources, had dwelled in the same hamlet since the fourteenth century, was suddenly obliged to abandon the place it had lived in for more than five hundred years and go away, separated from each other, toward the farthest corners of the earth. The two brothers left in Costa set off pretty soon afterwards, certainly before June 3 of the same year, because on that day Giacomo's name was included in the register of draft evaders, who had failed to turn up for their military service. And so it was that the parents, together with their youngest daughter and son were on their way to

Uruguay, the twenty-seven year old Maria Anna was about to leave for Brazil with her husband and three children, the twenty-two year old Giovanni was working in North America on the border with Canada, and twenty-four year old Francesco was with twenty-year old Giacomo in France, where Italian emigrants were the butt of increasing scorn and persecution. The documents in our hands regarding this aspect do not describe their situation in great detail, but it is not hard to imagine how much fatigue, anguish and pain the choices made and the pilgrimages undertaken cost all these family members.

## NOTES

1 The arrival of Giovanni Dalla Costa on the Pacific coast, in 1888, is recorded in the article "Jack Costa," in *The Tanana Gold Fields*, a single issue, Fairbanks, 1903, p. 19, Rasmuson Library Archives, Cecil Robe Collection, Alaska and Polar Regions Department, University of Alaska, Fairbanks.

2 This information is contained in a letter from Gilberte Arnoux to Mariarosa Dalla Costa, Figanières, dated December 30, 2004, in which, however, it is unclear (she writes "there was no money left") whether Teresa had used up money that had been deposited, whether the bank itself had gone bust, or whether Teresa had gone there to ask for a loan. The most likely version is that a request was made for a loan: if she had had funds deposited in the bank, Teresa would have known it… and it is highly unlikely that the house burnt down at the same time that the bank went bust: although there was no shortage of misfortune in those times, two huge disasters like these could hardly have occurred simultaneously…. The same letter also relates that for this trip of hers to the bank of Montebelluna, Teresa went in a gig and was wearing her jewelry. As made clear in the recollections of Giacomo Dalla Costa, it was well known to his descendants in France that Teresa loved dressing up in her jewelry.

3 Also in Uruguay laws had been promulgated in favor of immigration for colonial ends. The law of 1880 promoted agriculture with the institution of a special fund destined to creating or encouraging colonization initiatives by private enterprises. The colonies had to be situated in state or private land that could be expropriated when the number of families intending to colonize them exceeded a hundred and there were no other public lands available. But the funds were so meager that they even precluded a real demarcation of the state lands. This plan was nothing more than the formal expression of an unrealizable policy. However, this law, thanks also to the new project set out in 1888 — a forerunner to the more complete and ambitious Uruguayan law for the regulation of immigration promulgated on 18

June 1890 — led to the arrival in Uruguay of about three thousand Italian families from Liguria, Piedmont, Lombardy, Marche, Veneto, Emilia and Tuscany in the late eighties; their travel expenses were covered in advance by the Uruguay State, which explicitly excluded emigrants from Naples and Sicily. The documents available do not reveal whether the Dalla Costa family was entitled to undertake this journey free of charge (Juan Antonio Oddone, *La politica e le immagini dell'emigrazione italiana in Uruguay, 1830–1930* [Policy and pictures of Italian emigration to Uruguay, 1830–1930], in Fernando J. Devoto *et al.*, *L'emigrazione italiana e la formazione dell'Uruguay moderno* [Italian emigration and the formation of modern Uruguay], Edizioni della Fondazione Giovanni Agnelli, Turin, 1993, pp. 95–97).

4 The passport issued to Luigi shows that he, head of the family and already 55 years of age, left Genoa on March 16, 1890 with the Rubattino shipping company, together with his wife Teresa, aged 52 years, their daughter Elisabetta, aged 13, and Gaspare, aged eight. The destination was Montevideo, Uruguay. The document was provided by Marly Dalla Costa, daughter of Osorio Giacomo, Gaspare's son. She still lives in Guaporé, Rio Grande do Sul: Thanks to her, we were able to reconstruct the history of the branch of the family that emigrated to Brazil.

5 *Pergamene antiche dei nobili trevigiani conti di Onigo (sec. XIII–XVIII)* [Antique parchment manuscripts of the noble Treviso counts of Onigo (13th to 18th centuries)], vol II, *La prima dominazione veneziana (1339–1365), Pergamene 117–381,* Cataloging by Gabriele Farronato, Antiga Edizioni, Cornuda (Treviso), 2001, p. 451.

6 All those found to be absent for whatever reason, including legal emigration, were recorded as draft evaders on the day of their twentieth birthday. State Archives of Treviso, Registers for military service at the Fondo Archivio Storico Comunale [Historical Municipal Archive Heritage].

7 Toward the end of the nineteenth century, in all areas housing emigrants of Italian origin, there were numerous demonstrations of hostility towards Italian emigrants. In particular, in 1882 in the Nimes area of France, there was a real hunt for Italians working in the countryside. With a tragic ritual, they were tied to a tree, tortured and abandoned after losing consciousness – often they died. In Marseille in 1883 there were ferocious chases through the tortuous streets surrounding the port. There were also attacks, with immigrants left dead or wounded, at Toulon, in Moselle, in the Bourgogne and Isère. The worst incident, however, occurred in 1893 in Aigues-Mortes, southern France, the geographical area where the Dalla Costa brothers also found work. "At Aigues-Mortes there were important salt-works; many Italians also worked there. The owners exploited the availability of foreigners in response to local workers' wage claims. The Italians, who did not realize that they were pawns in a game that damaged the local workers, simply

needed to work, and considered any wages acceptable, however low. The anger of the local population, at boiling point, erupted in a horrific way. More than 500 Frenchmen attacked the 'big house' made of stone and canes that the Italians had sought refuge in. Shouting 'A la corde les italiens!' 'Let's hang the Italians!', they leapt onto the roof and threw wood and stones into the interior. Shortly before midday, a furious crowd attacked those who tried to save themselves. The wounded and the fallen were killed with pitchforks and truncheons. Babies and children were among them. According to testimony collected by a journalist for *Corriere della Sera:* 'the flesh of the children was cut into shreds. Their tiny bodies were impaled and carried in triumph …'. A local newspaper wrote: 'Things of such a ferocious nature have never been seen before, not even in Africa …'. The toll was horrifying: there was talk of 50 dead, and 140 wounded being hurled into a canal (*La valigia dell'emigrante, prev. cit.,* pp. 59–60).

# 16
# In the White Nome Desert

Without doubt Giovanni suffered greatly on hearing his family's sad news. While he prayed for his loved ones, the determination to somehow see things through was growing within him; more than ever before he felt he needed to escape his present situation and to try his chances at something new, some more favorable opportunities. He was working in the mines to the south of Seattle, in the Tacoma district, probably at Carbonado, in one of the many mining areas in the state of Washington. One of these areas that was particularly famous lay to the east, at the border with Montana, where a census conducted in 1890 recorded that in the same year 3,993 people were living in the encampments at the foot of the canyons, and they had extracted minerals worth about 10 million dollars. Wardner, the largest city in the area had almost 4,000 inhabitants. In 1890 a railroad was built to link Spokane and Wallace and, in 1891, electric light was available. There were saloons and gambling houses and dancing halls; people organized baseball matches, boxing and horseracing events. There were also four miners' unions which, founded in the fall of 1890, merged into a single front on January 1, 1891. The unions found their reason for being in the behavior of the managers, who constantly tried to slash wages and cut down on spending for the already rudimentary safety measures provided; they were also hell bent on making huge profits from sales, and from single miners' lodgings, all of which were under the monopoly of the mining company. The objectives of the workers' organizations were to keep wages high, combat the commercial monopolies of companies, ensure that accidents were prevented, obtain basic medical assistance services for the sick and injured, and secure a decent burial for the dead. In winter 1890, a strike was called in an effort to bring wages back up to 3.5 dollars a day, which had already been paid in the mining area before 1887, only to be reduced by the mining companies. The strike was successful and, for more than a year, peace was restored in the mining fields.

However, things took a real turn for the worse in 1892, when the association of mine owners decreed that the pay for working a ten-hour day was to be fixed at 3.5 dollars for experienced miners, but reduced to 3 dollars for wagon workers and shovelers. The union retorted that pay was to be equal for all and, in July of the same year, a revolt broke out in the city of Gem. The oil-mill was blown up with dynamite and some people died; to restore order, six companies of the National Guard and the federal troops were called out.

Rumors of these happenings, although taking place some hundreds of kilometers away, must have reached the ears of Giovanni, too, but other events that would transform his life had already happened then. Above all, his brother Francesco had turned up at Carbonado; Giovanni had written to him on several occasions to tell him how things were going and, above all, to ask him to join him.[1] Francesco told him that Giacomo had not felt up to taking the ship to America, and had stayed to work in France, in an iron ore mine.[2]

In the course of the past four years, Giovanni had struck up new friendships with other young men with whom he shared the ambition to find an escape from a lifetime of toil for a salary that, at a meager 3 or 3.5 dollars a day, was far too tight. And it was with his brother Francesco, and these Italian friends Stefano Albasini (Steve Absenna) and Felice Pedroni (Felix Pedro) and the Welshman, Lewis Jones that, in the spring of 1892, he decided to leave on an expedition to search for minerals in Cariboo County, British Columbia, Canada.[3]

To cover some of the costs, they had made the customary collection of funds; Giovanni and his brother got money from the local Italian community, Felix Pedro from Mister Auguste Hanot and Lewis Jones from the Welsh. The money was to be paid back with interest if the mission proved successful.[4] From the start, however, Giovanni felt in his heart of hearts that this was not what he really wanted. True, he was in the company of his brother and friends, but visions continued to flash through his mind, visions evoked by the tales of those who had gone further up north, to mythical Alaska, where the search for gold was still the prerogative of those few who were determined to do their all. So, gradually his decision matured, and he communicated it to the others.

After giving up his share in the enterprise and saying good bye to his brother and friends while promising to keep in touch, Giovanni returned to Seattle, boarded ship and, after a long voyage, reached an almost unknown port, on the western coast of Alaska. Nome[5] was the name of this godforsaken place.[6] On disembarking, Giovanni realized that he had done the right thing. The territory was immense and almost empty — but about twenty gold diggers could be seen in the surroundings. Too many for Giovanni's liking. After making a few attempts at searching to master the techniques required rather than make a real find, he bought the necessary equipment, including a pickaxe, an axe, a saw, a spade, a shovel and sieve, snowshoes, rawhide sleeping bags, a fur-coat, a rifle, a sled and animals to pull it — six San Bernardo dogs. After also acquiring six reindeer and as much canned food as possible for himself and his dogs,[7] he set off together with a companion, a Swede8, who eventually disappeared during a snow storm, leaving his companion completely alone. Since Giovanni had to travel through the huge, icy Alaskan north, he decided to take advantage of the last stage of winter, when the trail on the Yukon — the large river that went inland toward the heart of the immense peninsula for a stretch of more than three thousand kilometers - was still iced up, and therefore passable.

## Notes

1 Recollections of Maria Ghidelli, wife of Francesco the son of Giovanni Dalla Costa, reporting on information given to her by her husband and her mother in law, Rosa Rostolis.

2 Letter from Gilberte Arnoux to Mariarosa Dalla Costa, Figanières, November 13, 2004.

3 Alex Hanot, *The Story of Felix Pedro*, in *Hanot Papers*, Rasmuson Library Archives, Alaska and Polar Regions Department, University of Alaska Fairbanks.

4 Alex Hanot, *op. cit.*

5 The strange name of this place is explained by the following anecdote, reported by L. Pellegrini in "La febbre dell'oro" [Gold fever], an article that appeared in *Domenica del Corriere* on September 12, 1965: when a navy officer exploring the Alaskan coast saw this stretch of coast, he wrote "Name?" on the map as if he was asking — or postponing any definition of - its name. But then the person who interpreted his notes did not bother to consider the details and, misspelling a little, wrote "Nome" down: the name stuck.

6 All his descendants agree that Giovanni Dalla Costa got to Nome in the year 1892. Also see the article by the journalist Lino Pellegrini: "Un trevigiano 'aprì' la corsa all'oro nel favoloso Yukon" [Man from Treviso 'started' the gold rush in fabulous Yukon], in *Il Giornale,* 21 November 1990, p. 4 and in *Il Corriere di Caracas,* n. 50, 26 December 1990, pp. 1–2; "After years of

courageous searching, the brothers Giovanni and Francesco Dalla Costa discovered a large gold deposit," in *Il Globo,* an Australian newspaper, year XVIII, no. 2, 3 January 1975, p 3; "Alone in the dark inferno," in *Il Gazzettino,* 30 October, 1980, p. 3; "Nuggets, wolves and . . .crimes," in *La Notte,* 11 August 1982, p. 6: "Honeymoon amongst wolves" in *Oggi* year XXX, no. 50, 11 December 1974, p. 93 ff; "The Italians and the gold rush," in *Storia illustrata,* no. 322, September 1984, pp. 106–110, "An Alaska that we ignore. Journey in the 'gold fever,' which Italy took part in," in *Qui Touring*, year XXVI, no. 3, March 1996, pp. 76–82. Cfr also Daniele Ferrazza, "All'ultima pepita. Da Pederobba all'Alaska" [To the last nugget. From Pederobba to Alaska], in *Il Mattino di Padova*, July 21, 1996, p. 13.

Two articles appeared in *The Tanana Gold Fields,* single issue, Fairbanks, 1903, p. 19, Rasmuson Library Archives, Cecil Robe Collection, Alaska and Polar Regions Dept., University of Alaska Fairbanks. One was entitled Jack Costa, *prev. cit.,* and the other, *Frank Costa*. In the former, 1888 is indicated as the year in which Giovanni reached the Pacific coast, whereas the other specifies 1894 as the year of his first arrival in Alaska, also confirmed in Alex Hanot, *op. cit.,* and in the twelfth census of the United States made in Circle City, Alaska, in May 1900 (Twelfth Census of the United States, Census of Alaska). There is no official record of Giovanni's arrival in Nome in 1892, but this may depend on the fact that the local newspapers, in which passenger arrival and departure lists for Nome appeared every year until 1905, were lost, as confirmed by Connie Bradbury, the Alaskan researcher contacted by us at Fairbanks.

[7] Recollections of Giovanni (Gianni) Dalla Costa from the stories of his father, Francesco first son of Jack Costa.

[8] Lino Pellegrini, "*Da solo nell'inferno buio*" [Alone in the dark inferno], *prev. cit.*

# The Origins

Costa: the hamlet in which Giovanni Dalla Costa was born, with the Asolo hills in the background.

Passport of Luigi, the Dalla Costa gold seeker brothers' father, issued in Genoa on March 14, 1890 and authorizing him and his wife, Teresa Morgantin, to leave the country for Montevideo in Uruguay, together with their children, Elisabetta and Gaspare.

# ALASKA

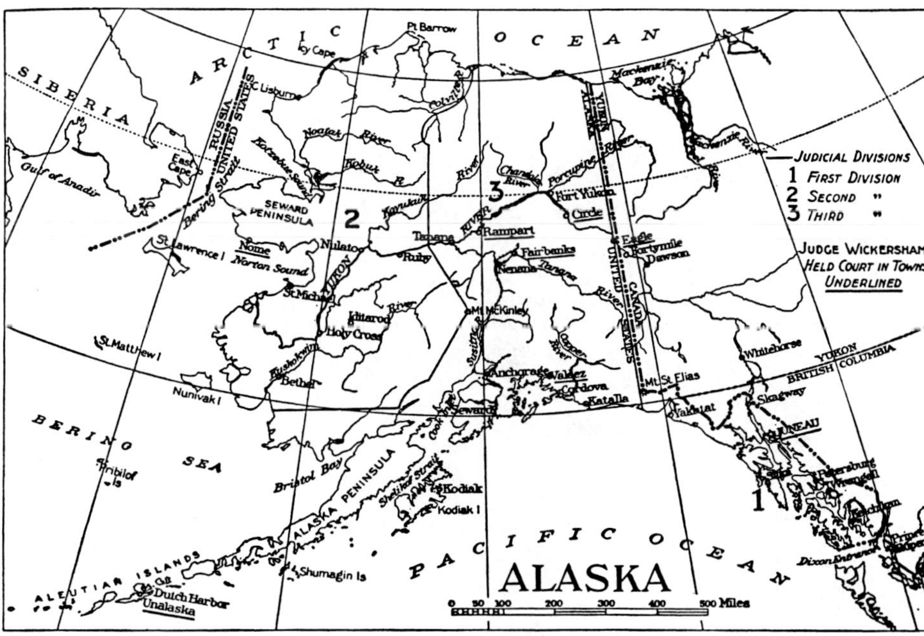


Map of Alaska from the book *Old Yukon* by Judge James Wickersham, who had jurisdiction in the locations underlined.

***(above)***

Francesco Dalla Costa and Felice Pedroni on an ox-drawn cart, Hot Springs, Arkansas.

***(left)***

Hot Springs, Arkansas: Francesco Dalla Costa and Felice Pedroni posing — tongue in cheek.

Francesco Dalla Costa, on horseback in Alaska.

# NEWS CLIPPINGS OF GENERAL INTEREST

### Carry Large Pokes.

While on his way to San Francisco for machinery, C. L. West was interviewed by a Skagway Guide reporter with the following result:

Mr. C. L. West arrived direct from Tanana, for which place he left Dawson last March.

Mr. West, when met by a Guide correspondent, was somewhat reluctant about speaking of the Tanana, as he does not desire to pose as a booster. Pressed for information, however, he said that the past two months had wrought a wonderful change in the Tanana country. Mr. West said, had he been able to get away from here about June 1, he no doubt would have given the camp a bad reputation, but that now he has no occasion to do so; that the country is being opened up and is showing good pay on all the creeks worked. Every man carries a good poke, and many of them are well filled.

On claim 2 above on Pedro, Costa brothers are working from twenty-five to thirty men, and their clean-up will compare favorably with those of some of the best paying claims in the Klondike. Within the past two months prospecting throughout the country has been extensively carried on; new creeks have been found in many places, giving to the Tanana what Mr. West considers the basis for a prosperous mining camp for many years to come.

### The Goodpasture Strike.

There is every reason to believe that the reported rich strike made late last fall somewhere in the Goodpasture region is proving good, although nothing definite is known about it yet. Five men on their way from Dawson to the southern slope of the Tanana, are supposed to have made the discovery. One of the party returned to Dawson early in the summer, with $2,000 in dust, and after securing a supply of provisions the party, with one intimate friend, returned to the hills.

The discovery was made by accident. While boating down some stream—supposed to be the Goodpasture—good looking wash was discovered on the rim of a high bench. A stop was made for dinner, and while the preparations were in progress, a wash-dish of the top gravel was tried, which yielded 34 cents. Camp was pitched, and further prospecting developed an extensive deposit of fabulously rich gravel.

Mining operations were at once commenced. Some of the men carried the gravel down from the bench in gunny sacks, while others washed it at the river bank with a rocker. In this crude way several thousand dollars were cleaned up before winter, when development work was vigorously prosecuted.

Dr. Medill was indirectly interested in the discovery, and was invited to join the miner on his return to the find; but as all his arrangements had been completed to locate in Fairbanks, the proposition could not be entertained.

The location of the stream is yet but a conjecture. It is supposed to be on a tributary on the Goodpasture, which flows into the Tanana about 140 miles above Fairbanks.

### Stampede Now On.

A party of three men from Chenoa, composed of Messrs. Sullivan, Smith and Harry Kato, a Jap, returned from an extended prospecting trip in the foothills of the Alaskan range of mountains, which form the southern wall of the Tanana. On the eve of their departure for home they discovered an unexplored section that they have every reason to believe will prove on further investigation to be rich. Their own research was prolonged to the limit of their supply of provisions, which was nearly exhausted when the locality was reached.

The story of the strike, so far as the News has been able to learn, is as follows: About seventy miles southeast from Fairbanks, on an unnamed creek tributary to the Tanana, the prospectors found wash that yielded six cents to the pan in the grass roots. As a result of the pannings $14 was brought to town. Bob Chamberlain of Fairbanks, who grub-staked the party, returned at once with them to the locality for the purpose of making further investigation.

Even the meager news that found its way to the public ear created considerable excitement, and several stampede parties were formed, only to meet an untimely end in the incipient stage, for want of information.

A party of about twenty, however, composed of men from both Fairbanks and Chenoa, took the trail in the hope that they might follow it in.

The discoverers only located one claim each for themselves and one for Mr. Chamberlain on each of the creeks discovered.

The first discovery was christened Japan Creek. Chenoa Creek and Golden Gulch were staked and named next in their order, and no further search was made. The gold exhibited was coarse, and showed but little wash.

DAVE CASCADEN

COSTA BROTHERS

### Burn Cabin for Signal.

As the steamer Koyukuk was preparing to leave her wharf at Battles on her trip to Fairbanks, her captain was approached by a party who requested him to delay departure for a while, as they were waiting for a party of two on the creeks who were desirous of leaving for Fairbanks. The captain said it was impossible to delay, but that he expected to be hung up on a bar four miles below town, and suggested that if the parties came in they would see the smoke from the steamer, and they could signal him by setting fire to a cabin and he would await their coming. The incident is suggestive of the feeling of all who are leaving the Koyukuk. The working out of a placer district is always sure to place the district town among the number of deserted towns, such as are found in Nevada and Idaho. Battles is now among the "has beens."

Gus Gulliksen, in company with William Cheesman, has a two years' lay on 5 below, Fairbanks.

Mr. Corbisher, who has the contract to put up the relay stations along the telegraph line from Tanana to Eagle, is pushing the work with all possible vigor. The stations are twenty miles apart.

Jack Turner, Tom Beatty, Phil Wren and Chester Johnson, who worked a lay on Wolf Creek this summer, are in town this week outfitting for the winter. They will work another lay this winter.

D. Ryan of Skagway has opened a cigar store on Front street.

William Prendergast, John McNeff and Mike Wendell, who have been working a two-thirds lay of 2 above on Pedro, made a good clean-up last week. After a three-days' run, with five men shoveling in, $1,800 was taken out of the boxes.—News.

The Dalla Costa brothers (picture on right) in Alaska. Francesco, seated, and Giovanni standing behind him with his dog in a photograph that appeared in the newspaper *The Tanana Gold Fields*, single issue, Fairbanks, 1903.

THE TANANA GOLD FIELDS 15

### NEW DISCOVERY.

**Big Quartz District Found Across the Tanana.**

Tim Crowley, Jack Pomeroy and Sam Barber returned lately from an extended visit to the new quartz district discovered by them last summer. The district is on the headwaters of the Black River, in the Alaskan range of mountains, and is distant from Fairbanks in a direct line, almost due south, about 100 miles, but by the route traveled is nearly 200 miles.

They located two monster ledges, one of which is 100 feet between walls and crops 100 feet high for fully

JACK COSTA

half a mile. There are many ledges there, and float quartz can be found everywhere. They were not able to judge the character or probable value of the mineral, but think it to be a high grade silver ore, carrying a large per cent. of copper. A number of samples were brought out for assaying. Some members of the party will return before spring to spend the summer in developing the property. But few locations have been made there yet.

### FAIRBANKS' FIRST SOCIAL FUNCTION.

The first social function given in Fairbanks took place Thanksgiving night in the form of an invitation ball given under the auspices of the Arctic Brotherhood. The event was greatly enjoyed by all who were the lucky participants. The dancing was in the A. B. Hall, which was decorated tastefully and festooned in rustic style. At eleven o'clock there was a grand supper march, and the gay party repaired to a building opposite, which had been specially fitted up for the collation. The banquet hall was tastefully decorated with large American flags and fir boughs. There were two long tables the full length of the room, and one along the end. The tables were neatly spread and filled with all the good things which go in keeping with a Thanksgiving dinner. The following were among those present: Mr. and Mrs. Kellum; Mr. and Mrs. Turner, Mrs. Charles Hill, Mrs. Fassel, Miss Barrett, Mrs. Frank Wilson, Mrs. Jessie Evans, Mr. and Mrs. C. Nichols, Mr. and Mrs. Dupras, Mrs. Mitchell, Mrs. Leonard, Mrs. Werner, Mrs. Gibson, Mrs. Davis, Mr. and Mrs. Hyde, Mr. and Mrs. John Rapp, Mrs. Delaney, Messrs. Dodge, Condon, Spring, Solly Spring, Overgaard, Noonan, Hall, W. H. Smith, R. D. Smith, St. George, Meehan, Cleary, Laymon, McDonough, Fox, Dupras, Long, Karsden, McGonigle, Bennett, Wilson, John L. Long, Wiedman, Gamble, Bellows, Carter, Lindig, Sherman, Healy, Wood, Petree, Henshaw, Fain, Parker, Woitke, Mogean, Dr. Hall, Cassely, Harrington, Robertson, Miller, Judge Cowles, Crede, Woitke, Siebe Hoffman, Taylor, Petree, Rockwell, Wachter, Deane, Kinman, Parker, Shaw, George Mutchler, Purchase, Dr. Ehie, Dr. Danforth, Oldfield, Sergeant Murphy, Sergeant Smith and Corporal Martin.

The Fairbanks News office is prepared to fill any order for fine job printing.

**Telephone for Fairbanks.**

The Yukon Telephone Syndicate, with Mr. N. A. Fuller, president, will establish a telephone system at Fairbanks. The material is now on the way. The capital of the company is $50,000.—Dawson Daily News.

**Express for the Creeks.**

The Arrow Express leaves Fairbanks for the creeks every Sunday at ten A. M., carrying mail, papers and small packages. The rates are 25 cents for letters or papers, and 50 cents for small packages up to 5 pounds. From 5 to 10 pounds, $1. Anything above that, 7 cents per pound. Terms cash. Leave orders at News office.

**Tanana Turnips.**

George Harrington of the Fairbanks has a fine garden at Chena. Though the growing season is not yet over he has turnips that will weigh 30 pounds, cabbage as large as water buckets, beets as good as can be grown anywhere, potatoes fit for the gods, lettuce in abundance, and kale is high as Deacon Jones's fence. There is little reason why our vegetable supply can not be home grown.

**Important Decision.**

In the commissioner's court Judge Cowles decided the trespass case of L. Meyers vs. Johnson & Meredith in favor of the plaintiffs. The gist of Judge Cowles' decision was as follows: Every man has a right to relocate mining ground, stake fractions or otherwise contest an original locator's right to mining ground, but such subsequent locator must stop here, and before he takes possession of the ground and proceeds to work same must establish his right to same in the district court; until that time he is a trespasser.

**Al Morency Arrives.**

Al Morency, one of the best-known miners in Alaska, arrived lately from Circle, and he intends putting in the winter here, which is sufficient evidence of his good opinion of this part of the country. Mr. Morency has been mining on Mastodon, and on other creeks in the Birch Creek country, since '94, and was also a heavy operator in the Klondike country.

It is roughly estimated that Al has taken enough gold out of the ground to load a mule pack-train forty miles long, and if the gold had been coined into $20-gold pieces, and were placed edge on edge they would make a string long enough to reach within a very few miles of the moon. Al has a reputation as a phenomenal miner; he was never known to work a piece of ground that would not pay. It is said of him that he can tell whether ground will pay to work before ever a pick has been stuck into it.

Those who know him best say that it takes just $94,000 to make a comfortable grub stake to last him through an Arctic winter, and even then he has been known to do a little rocking early in the season before the ground had thawed sufficiently to get a string of boxes in operation, in order to paint a parting streak of carmine around the rough edges of old Circle.

**Mining Notes.**

H. L. Meyers of 3 Chatham rocked out $600 in five days.

On 3 above on Fairbanks Creek they have been getting $2 to the pan.

Tom Rockwell and [illegible] have taken a 100 foot lay on No. 4 below, Fairbanks. They will take out a big dump this winter.

Two new roadhouses have opened up lately. One is the Owl House, about nine miles out of town, and the other one is on No. 5 Gold Stream.

The Gold City roadhouse, built by H. Howard, contains all the accommodations for travelers and wayfarers. There is in connection a good stable and doghouse.

Louie Laflames, well-known the country wide as the owner of large mining interests on Mastodon and other creeks, arrived from the Circle district. He is the owner of 6 and 7 above on Pedro, where he will take out large dumps this winter.

Sam Blackburn, James and Jack McPike of Independence Creek arrived in town Monday. These gentlemen are the owners of one-half interest in 1 below on Fairbanks, where they will take out a large dump this winter.

### TELEGRAPH FACILITIES.

Fairbanks is in direct communication with all the rest of the world. The United States government telegraph line extending from there to Dawson connects with the telegraph line of the Canadian government; from thence the telegram travels by way of Ashcroft, B. C., and from there on the messages are mostly handled by the Western Union Telegraph Company. The rates for messages from Fairbanks to cities along the coast is about $5.50 for ten words.

The Marconi system has been installed along the Tanana River from Fort Gibbon to Chena. So far the workmen engaged in adjusting the wireless system of telegraph have been unable to accomplish anything like satisfactory results. The uneven nature of the country is a serious bar to the successful working of the system, especially where powerful machinery has not been installed to generate sufficient electrical energy to attract the deflected wave.

### ITEMS OF GENERAL INTEREST

A 25-horse-power boiler, a 4 and a 6-inch centrifugal pump, and a steam saw have been installed on 3 below, Fairbanks.

John L. Long is the mining recorder for Fairbanks district. Mr. Long is also the clerk of the court for Fairbanks, and a member of the city council, having received the largest number of votes cast in the recent election.

Messrs. Rossiand, Moreau and Deschamps, who have been working a lay on Discovery, Chatham Creek, arrived in town Thursday with a fine clean-up. The last ten days, from September 5 to 15, with one man shoveling in, 34 ounces was cleaned up.—News.

A contract has been let for the erection of a two-story hotel on the Butler & Gibson lot, just above the Commissioner's office, on First avenue. The dimensions of the structure will be 30x60 feet, and it is the intention of the owners to have a first-class hostelry in every particular.

Between forty and fifty buildings are now in course of construction, and others are being started daily. Fairbanks is the largest log cabin town in the world to-day, and the number of buildings will soon be doubled. There are but few vacant buildings, those now under construction are not for rent, and the population of the town is rapidly increasing.

DICK MOTSCHMAN

The preliminary survey for the proposed railroad is now under way. The promoters have run a survey line along the streets of Chena. They expect to run the survey to Pedro Creek, and announce that if the business on the creeks warrants they will put in a road during the coming year.

Giovanni Dalla Costa in Alaska (picture on left) in an article published in the newspaper *The Tanana Gold Fields*, single issue, Fairbanks, 1903.

The steamer *Lavelle Young* mooring at Fairbanks, Alaska.

Stake no. 2 above on Pedro, which made Giovanni Dalla Costa a rich man.

The sluice, in action.

Giovanni and Rosina in front of the mine;
behind them, on the right, Francesco Dalla Costa.

Jack Costa in Alaska with his wife Rosa, brother Frank, two children and his dog, Jocko.

Jack making a toast with Rosa, Frank and others in front of the log cabin in Alaska.

Francesco Dalla Costa in Alaska with his wife Maria and two friends.

Francesco Dalla Costa with his wife Maria in Alaska seated in front of their wooden house.

Maria Polesellli, wife of Francesco Dalla Costa, sitting in front of her wooden house in Alaska cradling her newborn baby, Rena.

# BRAZIL

Gaspare Dalla Costa with his wife,
Virginia Gotardo, and their first three children.

Guaporé, Rio Grande do Sul: the tailor's shop (foreground) run by Ervino, Gaspare Dalla Costa's son, with Hotel Roma just behind it.

***(above)***
Guaporé: The hotel owned by Gaspare Dalla Costa at ground level, with its bar (left) and restaurant (right).

***(right)***
Guaporé: The tailor's shop owned by Gaspare and run by his son, Ervino.

***(above)*** Gaspare Dalla Costa with his wife Virginia and children Rosalia, Aurelio, Genoveva, Osorio Giacomo, Germano Giacomo, Lidia, Ervino and Gentile.

***(middle)*** Gaspare and Virginia Dalla Costa in Hotel Roma, behind the bar.

***(below)*** Gaspare Dalla Costa and some guests outside Hotel Roma.

## France

The steamship *Duca di Genova* in which Giacomo and his family reached Brazil in January 1915.

Giacomo Dalla Costa with his wife
Marie Granet when living in Brazil.

Giacomo Dalla Costa.

Clotilde and Blanche,
Giacomo Dalla Costa's daughters, in Brazil.

Figanières (Provence):
The house that belonged
to Giacomo Dalla Costa.

Figanières (Provence):
Recent photograph
of the bar run by
Giacomo Dalla Costa
when he returned
to France from Brazil.

# Pederobba and Giovanni's Return

Portrait of Giovanni Dalla Costa with his wife Rosa.

Giovanni Dalla Costa with his wife and children:
Margherita (Rita) on her mother's knee, Marianna (Bettina) standing, Teresa (Resi) seated, and Francesco.

Francesco Dalla Costa,
firstborn son of Giovanni,
at one year of age.

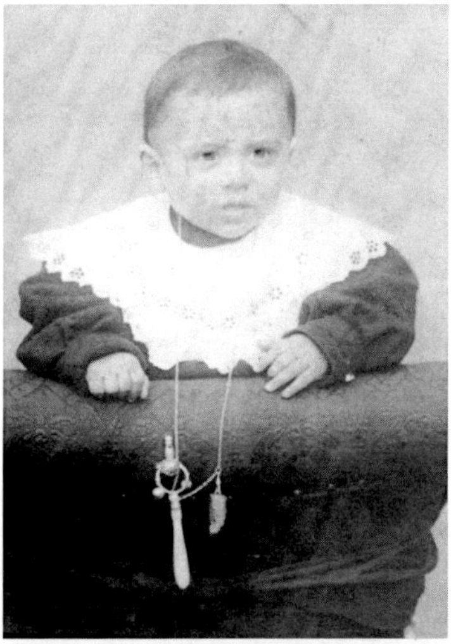

Francesco,
firstborn son
of Giovanni
Dalla Costa,
with his sisters:
to the left
Marianna
(Bettina),
to the right
Teresa
(Resi) and,
on his knee,
Margherita
(Rita).

House of Giovanni Dalla Costa at Pederobba
as seen from the main street, Via Roma.

Teresa (Resi), Giovanni and Rosa's daughter, who died of Spanish 'flu at Pavia.

Francesco, son of Giovanni Dalla Costa, on his beloved white horse.

Giovanni in a one-horse carriage, surrounded by his family,
in the courtyard of the house at Pederobba.
At the top of the steps, the house-servant Giustina Bresolin
and the coachman Giacomo Bresolin, nicknamed Metto.

The memorial slab on the family tomb of
Giovanni Dalla Costa in Pederobba cemetery.

# Pederobba and Francesco's Return

*La Provence,* the transatlantic ship that Francesco Dalla Costa and his wife Maria sailed on for their honeymoon.

Francesco Dalla Costa with his wife
Maria and their four daughters.

Francesco Dalla Costa with his wife Maria Poleselli.

Francesco Dalla Costa with his wife Maria Poleselli
and daughters Rena, Lidia, Ester and Alice.

Francesco Dalla Costa's house at Pederobba.

Villa Fairbanks in Treviso, which Francesco Dalla Costa planned and had built in 1928 for himself and his family.

## Keepsakes

The small trunk that Giovanni Dalla Costa buried in the ground.

Pendent and brooch that Giovanni Dalla Costa gave his wife Rosa.

The gold fob-watch that Giovanni always kept with him.

***(left)*** The brooch decorated with tools used by gold diggers.
***(right)*** Giovanni Dalla Costa's watch chain made out of gold nuggets.

***(left)*** Piece of jewelry made with the first gold nugget found by the Dalla Costa brothers.
***(right)*** Family pendent: a spade with a gold nugget.

Part of the necklace made of gold nuggets given to Rosa Rostolis by Giovanni Dalla Costa.

# 17
# Alaska

Where would the adventure of Giovanni Dalla Costa lead him? What was Alaska like in those times? To answer this question more fully we must go right back to the end of the eighteenth century, to Russia, when Peter the Great was Tsar. Until then, the great stretches of Siberia had been an apparently inexhaustible source of animals for fur, but year by year the precious animals were getting scarce due to indiscriminate hunting. The Tsar therefore ordered hunters to push further to the East, not only in the search for new and richer hunting areas, but also to see whether there was a land passage to the Americas.

Peter the Great died in 1725 — without getting any answer to this question. Another sixteen years were to go by before the Dane, Vitus Bering, explored the area and discovered the strait that was to take his name, and the territory that extended beyond that arm of sea, a boundless virgin world that the few natives living there called Alyaska, the Great Land. From then on, for more than a century, the Russian fur hunters rampaged this vast virgin territory. Apparently some of them had realized that gold was to be found there, but they considered this of little importance compared to the incredible bounty of bears, foxes, mink, martens, ermines and beavers, which provided them with easy and sure money.

The fate of Alaska changed when Russia decided to get rid of this distant region, deemed unproductive because the supply of animals for fur had been exhausted. In 1867, it was sold off to the United States for just over seven million dollars.[1]

If one were to search for analogies to the situation of Alaska after its acquisition by America, many would be found in post-1848 California. The largest of the United States was an infinite stretch of mountains, valleys, rivers, woods and forests. Practically uninhabited, it needed to be colonized and developed from scratch. Its evident qualities were its intact, albeit harsh, nature, with plenty of varied and diversified fauna — and a climate that for most of the year had temperatures that dropped to forty and fifty degrees below zero. In spite of the tremendous efforts made by the Russian hunters, there were indeed still plenty of healthy animals of all sorts in Alaska. Above all there were bears, most of which were docile, but the Kodiak and grizzly subspecies were potentially lethal, even to armed men. The rivers and torrents were the kingdom of the intelligent and industrious beavers, which shared the pure waters with elegant otters, trout and salmon. Musk rats roamed free, as did wolverines and rabbits, and there were so many birds in the skyway that their songs could even be heard in winter. It was also common to come across herds of moose, caribou, mouflon, and white snow goats. And there were plenty of those legendary grey Alaskan wolves, always in packs, and extremely frightening for travelers because they had mastered insidious and sly attacking strategies, the aim usually being to plunder men's supplies and kill their dogs rather than to kill their human masters. But in this setting, these attacks were invariably fatal for the unfortunate victims.

So this was Alaska, a world in which it was an effort even to breathe: at minus 35° the air was "sucked in" like icy water, at minus 45° your breath would freeze as soon as left your mouth, and between minus 50° and minus 55° it had already frozen in your mouth, making breathing noisy and difficult; at below minus 55°, your spit froze before it reached the ground. To come here and search for fortune involved superhuman efforts and total isolation, the will and courage to face a thousand difficulties and to risk serious diseases; moreover, in summer the sun shone the whole time, but in winter there were only two or three hours of daylight, and for the rest of the twenty-four hours it was dark. This is probably why the gold rush did not have the same impetus in this area as it did in California. Since 1881, people had known that fair quantities of gold lay under the hard, frozen Alaskan earth, but the rush was certainly taking its time to gain momentum.

In 1885, the mining chronicles recounted that gold had been found on Fortymile River and, in 1892, it was found on Birch River near Circle City. This news must have also reached the ears of Giovanni Dalla Costa who, stepping into the white world of the Valley of the Yukon, reached these areas and began to search them. How did one look for gold in Alaska? Above all it was indispensable to have a parka, a large blue shirt of rough cotton with a fur bordered hood going down to the knees without any opening either in front or behind; this was put on over the head. What was the parka for? This simple garment worked as an effective barrier against ice. On entering a closed environment, its wearer took it off and left it to thaw. As soon as the wearer went outside again with it on, his parka would freeze up on the spot, but the clothes worn underneath it were still warm and dry. Another must for transporting equipment and provisions was a sled drawn by dogs, usually Saint Bernards, collies, mastiffs and, above all, malamutes, the sturdy Eskimo dogs that looked like wolves, and thrived in the harshest ever climate.

The equipment required for the search was practically the same as that used for the gold rush in California: the shallow pan, the rockers, sluices and the Long Tom. But there were some fundamental differences between this and the Californian search, for the latter could be undertaken undisturbed all year round whereas in Alaska the first and foremost problem was the land, which was always icy, thus preventing men from directly digging. First the ground had to be thawed with large wood fires which were relit day by day until the underlying earth was softened. Only after this treatment could the digging start. The next really tricky problem was water which, in Alaska, was available in large quantities only in the summer months whereas in winter it turned into ice. And we already know how much running water was needed for extracting gold from the gold-containing matter. Gold seekers in Alaska therefore had to work in a very short space of time, about four months from the end of May to the end of September, during which checks had to be made to work out which area was the richest in gold. How was this gold prospecting undertaken? After filling a large shallow pan with sand and gravel containing gold, the miners submerged the filled pan in water and then lifted it out, and gently swirled it using wide arm movements so that the water and lighter materials slopped out of the pan, and the gold settled at its base. The dense mush left behind was checked: if a quantity of gold that could fetch only ten cents was found, then it was useless to continue to dig in that spot. If fifteen cents' worth was found, then things were more hopeful, and it was advisable to continue digging. If twenty cents' worth of gold was found, then the seekers were on the right track for the Bonanza. But when the water froze, nothing could be done except to start digging without even having the chance to see whether the earth being dug up was rich or poor in gold. Of course a little water could be obtained by heating up the ice so as to make checks to work out whether the digging was being done in the right direction, but only in the following spring, when the land thawed, could the final evaluation be made to establish whether all the work done throughout the winter had been worth it, or just a waste of time.

These brave men, who crossed the ice on sleds and navigated rivers in rowing boats, had to face countless difficulties, suffering the harsh cold in the long winters and attacks from the tundra mosquitoes in the summer, moving across endless swamps without pathways or landmarks. These original adventurers were explorers true and proper; they were geographers in spite of themselves, being committed to traveling and discovering, charting maps and routes, creating trading points and rudimentary services in a universe where, until their arrival, the only men to have explored it were the hunters from Russia.

In this harsh environment, Giovanni Dalla Costa, twenty-four years of age and with a robust physique, one meter ninety-three centimeters tall and weighing one hundred four kilograms of bone and muscle tempered by hard physical work,[2] stood out among this handful of unknown heroes, strong men committed to the last frontier of the mythical gold rush. Throughout the days without light and the nights without dark of the Artic Pole, the youth underwent a slow metamorphosis until he became Jack Costa, a sourdough true and proper, a veteran seeker. He dug and dug in the Forty Mile area without coming up with anything much; then, in fall, he pushed onward until he reached Circle City with a view to checking out the land by Mastodon Fork and Miller Fork. The exploration was to be achieved bit by bit; victuals were rationed and a distance that could be covered in the harshest possible weather, was always kept with the nearest center for stocking up on supplies. Like the other seekers, Giovanni left all the gold that he found deposited in the few banks present in the area.

It was a hard and lonely life and, although he liked it, he began to long for the company of the others, whom he had left behind on the way to Cariboo County. So, in fall 1893, he decided to return to Carbonado where he was warmly welcomed by his brother and his friends, who had just returned empty-handed from their expedition in Canada — and he met a new companion, an Italian, Espalmer, who, like him, was an Alaska "veteran." Together with this group he agreed to set off again for the Yukon territory.

All together, apart from Francesco who had decided to stay in Carbonado a while longer, the group started out toward the end of February 1894. They traveled by boat to Juneau, and then had to cope with the legendary Chilkoot Pass, being among the first ever to travel along a trail that, as we will see further on, many followed — each carrying his own baggage of hardship and indescribable suffering.[3] After a difficult and adventurous journey, in the summer of 1894, they happily reached Forty Mile where they split up, each going his own way, to dig and sieve in the hunt for a fortune. Giovanni spent another two years of this life of solitude, of days and nights alone, of cold and fatigue. It must have been a relief when he heard that his brother Francesco had finally reached Alaska. From then on the destinies of the two brothers from Treviso, Giovanni (alias Jack, and sometimes John) and Francesco (alias Frank) Dalla Costa, were to be interwoven for long years in the infinite Alaskan universe: they were often together, and sometimes apart.

It was spring 1896[4] and, at last, Jack could count on help from his brother. He was no longer alone in the Alaskan gold rush.

But what gold rush? Wasn't Alaska only an icy desert crossed by a few steely individuals? Yes, it was still like this, but something was happening, and the wind of change could be felt in the air. This was when, after a series of promising and disappointing finds throughout Alaska, that the first really huge strike was made. In mid August, 1896 in the valley of Klondike, along the course of the river soon to be named Bonanza Creek, the amazing deposit of George Carmack[5] was found. The news spread like wildfire and, within a few days, all the claims along the river were picketed. They dug like madmen for an entire winter and, when the first spring warmth allowed them to sieve the pebbles and earth, the findings went well beyond the most rosy expectations and many found that, from one day to the next, they had become stinking rich.

## NOTES

1 For further information on emigration to Alaska as from this date, see F. Norris, *North to Alaska: An overview of Immigrants to Alaska, 1867–1945*, Alaska Historical Commission Studies on History, no. 121, June 1984.

2 Alex Hanot, *op. cit.*

3 Alex Hanot, *op. cit.*

4 The date of arrival of Francesco in Alaska is indicated as spring 1897 in *Frank Costa, prec. cit.* But the twelfth United States census undertaken in the mining district of Birch Creek from February to May, 1990 specifically states that he arrived there in March 1896 (Twelfth Census of the United States. Census of Alaska).

5 W. Greever, *op. cit.*, vol. II, pp. 193–197.

# 18
# Klondike

It was this way that, on the morning of 15 July 1897, the bells rang out through the whole of San Francisco. What was happening? A fire? No! It was gold! There were more than six hundred thousand dollars on the Excelsior, which was just being anchored! Masses of people were going down to the port to witness the event. There, they saw coming off the shabby old vessel those who had been the first to make a fortune in Klondike: wrapped in their heavy threadbare clothes, with wide brimmed hats well on their heads, long beards and weather beaten faces, they looked like a gang of tramps compared with the well turned out and refined city inhabitants. But in their bags, in their sacks, in their knapsacks, in cases and in any other possible container, they had stuffed so much gold that its weight called for the strengthening of the ship's exterior to ensure that its girders would hold. And a couple of days later exactly the same scene was repeated with the arrival of the Portland. Following these fantastic events, but also because of the economic depression that only a few weeks earlier had brought many companies in the United States to their knees, a new disease soon to be called "klondikitis"[1] broke out, not only in this city, but throughout the entire western coast of the United States, in Tacoma, in Portland and Victoria — above all in Seattle.

The disease was fulminating — there were stories of barbers who left their clients half-shaven, of owners leaving their stores without bothering to lock them up, and of preachers who ran off with their tunics billowing out behind them.

The lust for gold was so compelling that even important officials left their practices and seals to go in search of the windfall in one of the most distant and remote corners of the earth. Many of the ships in the Pacific Ocean had been moved to this coast to be ready to sail north. Steamships, cutters, transport and fishing boats, anything that floated and could somehow carry people, equipment and animals, was employed for the enterprise. By August 5, the Queen, the Mexico and the City of Topeca had set sail, followed hard on their heels by the Williamette, the City of Kingston and the City of Seattle,[2] and so on, in an impressive crescendo, fuelled also by the press, by the newspapers and the journals that, also to distract the public's attention from the difficult economic situation, did nothing but talk about gold, about the Bonanza and the magical opportunities open to all in the country of the midnight sun. Some more realistic opinions were also expressed: one newspaper in Great Falls, Montana, published an article entitled "Thousands of people famished in Klondike" and in the sub-heading "Almost two thousand graves in three years. The strain is unbearable." These sporadic warnings did little to stem the mounting tide of bounty hunters, but certainly few of these prospectors could ever have envisaged the life of toil, danger, disappointment and deprivation that awaited them in that wild land.

The first terrible taste, which often made the less determined give up, was waiting for them right at the beginning of the venture: the journey to Klondike that, after transferring onto a ship up to Juneau or Skagway, took an extremely heavy toll in terms of courage and physical resistance. The most commonly used routes in the rush of 1897 were Chilkhoot Pass and White Pass. Both started within North American territory, soon to reach the border with Canada. The remaining part of the journey was made across Canadian land up to Dawson and Klondike, both regions straddling Canadian and US Alaskan territories.

The first route, better for men without draft animals, left from Dyea, an encampment about three miles[3] away from Skagway. This trail was for paupers, who possessed nothing but their heavy equipment, their legs, and the will to throw their hearts past any obstacles to reach the longed for goal. And if already in 1897, notwithstanding improvements undertaken to make the route practicable by thousands of men that were reaching here from all over, it still was a really hard enterprise, it had certainly been much harder and wilder three years before, in 1894, when Giovanni Dalla Costa pressed along it with his companions.

For the first six miles, the trail followed the river Dyea upstream, flanking the mountains. After another thirteen miles, along a track made more impervious by a series of passes that were quite difficult and risky, one reached the so-called Sheep Camp, a settlement where, during the gold rush boom, there were rarely less than 1,500 people; it had taken on this name because it had long been the traditional base-camp for mountain sheep hunters. Now it was a camp for the gold prospectors, a place that thousands of people reached, already tired out, only to realize, in terrible winds and snowstorms, that there was worse to come: Chilkhoot Pass, given this name by the Chilkhoot, the Indian tribe that had always lived there. The best season for coping with it was winter, when the pass was like a series of shelves of ice upon which men tentatively traveled forward, making headway step by perilous step. In the four summer months, on the other hand, it was transformed into an impracticable slope, muddy and slippery with gravel, where every step forward was almost invariably followed by an equal or greater backslide. So, it was better to brave it in winter, notwithstanding the merciless cold and the constant danger of terrible avalanches, like that of April 27, 1898, which killed sixty people.[4]

After leaving Sheep Camp, the journey continued for almost three miles along a trail that leading to a wide gorge that was to eventually narrow into a funnel, closing in on the start of the pass itself. From here the journey was continued, travelers climbing, crawling on all fours on the ice, passing drops of thirty to sixty per cent, first up a path and then across the Chilkhoot glacier, right up to the top.

On arriving at this height one was faced with an infinite panorama: in the foreground was the large glacier that, descending along the opposite slope, gave birth with its waters to Crater Lake and then, further beyond, was a deep valley, crowned on both sides by high mountains. In the background, far away, lay two other large mountain chains, stretching as far as the eye could see. This magical view was overwhelming, but it also inspired anxiety and fear: "Where exactly should we go," these worried men would ask, " to find the gold?" And they were right to be concerned. After all, their journey was only just beginning. To get to the gold-bearing region they still had to overcome Turtle Hill, tackle the deep Tuci valley, and then press on to the shores of the great Lake Bennet, which was still more than a thousand kilometers away from Klondike valley.

The other route, as mentioned above, was the White Pass trail; almost eighty kilometers long, it was more feasible for travelers with horses, oxen or dogs for transporting their supplies and equipment. One left direct from Skagway, along a path that, for about seven miles, was wide enough to allow carts to pass through. But then the pass bottlenecked and, in some points, became a mule and horse track measuring less than a meter wide; overhanging it along one side were the steep grey walls of Devil's Mount and flanking it on the other was an almost sheer drop down a hundred-fifty-meter-deep crevice spangled with sharp rocks.[5] This route, due to its ground conditions and the cruelty of the men who used it, soon took on the sinister name of "dead horse trail." There were animal carcasses all along it.

The pass was often closed so that it could be cleared with dynamite and shovels. A farrier, who had set up a makeshift but useful smithy along the route, recounted that people had him shoe horses with bleeding hooves and legs. Many of these animals had pus underneath their pack-saddles, and their hide was ulcerated by scratches and abrasions. The gold seekers crossing White Pass had no pity for these beasts. Enormous packs of provisions were tied onto the pack saddles, and spurs were thrust deep and mercilessly into their flanks; when the horses were wounded by sharp rocks, fell into muddy trenches or simply buckled under the overload, they were often abandoned without their owners even having the compassion to finish them off with a bullet. Horror lurked behind every corner, and no less than three thousand animals died on this terrible slope, the eyes of the carcasses were gouged out by crows, and their bodies lying in the rivers polluted the water, making it undrinkable.

Jack London, who, in the fall of 1897 was making his way toward Klondike right along this trail, wrote:

> ...[the horses] fell off the trail, what there was of it, or they went through it; in the river they drowned under their loads, or were smashed to pieces against the boulders; they snapped their legs in the crevices and broke their backs falling backwards with their packs; in the sloughs they sank from sight or smothered in the slime, and they were disemboweled in the bogs where corduroy logs turned end up in the mud; men shot them, worked them to death, and when they were gone, went back to the beach and bought more. Some did not bother to shoot them, stripping the saddles off and the shoes and leaving them where they fell. Their hearts turned to stone - those that did not break — and they became beasts, the men on the Dead Horse Trail.[6]

However, whether they had arrived after making the terrible effort required for Chilkhoot Pass or whether they survived the hellish White Pass, the gold seekers from the different routes eventually converged on the aforementioned Lake Bennett, which was nothing but a point of departure for traveling to Dawson and Klondike by water. To go down the current one had to wait for the spring thaw. In the meanwhile, the men made flat based, square shaped boats, wide at the stern, their edges jutting out, and little more than seven meters long and less than two meters wide. The timber for making them was collected from the vicinity and cut into planks that were nailed together and then sealed with fabric and pitch. With a thick mast and a large sail of waterproof material, the typical Lake Bennett boat could carry two or three men and a load of two to three tons.

The great regatta left at the end of May. Hundreds of sails slit the waters of the enormous Lake Bennett until they passed it, slowly descending to Lake Nares and then on to Lake Tagish and Lake Marsh. From this point onward, the problems started. Before Miles Canyon, into which the Lewes River flowed with a first, strong whirlpool, and then in sequence, the rapids of Squaw and those of White Horse, thunderously foamy for about five hundred meters between two rock walls that stood sixty to ninety meters high.

In 1898 ten people died along this tract, prompting the Canadian police, the famous red jackets that followed events day by day, to prohibit the passing of unsafe embarkations and to force women and children to do this part of the trail on foot. For the less proficient — or less bold — there were pilots ready to make the dangerous crossing on a boat for a sum of money. One of these was Jack London himself who, this way, earned the three thousand dollars that enabled him to reach Dawson with a good pocketful of cash.

From here on, still along the Lewes river and then on the picturesque and windy Lake Labarge, the journey was smoother. The last piece of excitement was the Five Fingers rapids, beyond which peaceful navigation was resumed right up to Dawson. Toward mid-summer 1898, the vessels anchored at Dawson — although lined up in rows of six — they formed a long compact line that was almost four kilometers long.[7]

## NOTES

[1] Thomas Jeier, *La corsa all'oro (Alasca 1896)* [The gold rush (Alaska 1896)], La Scuola, Brescia 1977, p. 25. We have kept the word "Alasca" as it appears in the title of the book.

[2] Thomas Jeier, *op. cit.*, p.26.

[3] The mile, an English and American unit of measurement, corresponds to 1,609.3 meters, unlike the international sea mile, which is 1,853.18 meters.

[4] Thomas Jeier, *op. cit.*, p.74.

[5] Thomas Jeier, *op. cit.*, p.53.

[6] Thomas Jeier, *op. cit.*, p.52.

[7] W. Greever, *op. cit.*, p.226.

# 19
# Gold Fever Melts the Ice

How did these people reach Klondike after traveling thousands of kilometers in an unknown and hostile environment? They had rushed away from the prairies of the United States, from California and also from Europe, with only one idea clearly fixed in their minds: finding gold. Most of them had no inkling of what awaited them in the kingdom of ice. Many had only brought the heaviest coat or jacket that they had with them, plus their most precious possessions. Conditions here, however, called for much more than this.

We can safely say that during their journey from Skagway, or Dyea, the seekers had learned a long, hard lesson on what was in store for them once they reached Dawson. They had traveled a tortuous and painful path of initiation to the secrets of the extreme North during which many gave up and others lost their lives — all of them had borne great hardship. Nevertheless, while in fall 1896 the population of Dawson counted only 500 souls, in the summer of 1898 the number had rocketed to 30,000. These tens of thousands of gold seekers, most of them unprepared, were in great need.

In the large city of tents that extended along the river, as in many settlements scattered on the surrounding hills, in the valleys and on the nearby mountains, in winter 1897 there was also a crisis in the food supplies, and this meant that, in addition to all the other adversities, the entire settlement suffered from hunger. And, during this tremendous winter, Giovanni Dalla Costa, who had abandoned his search for gold for a time since it had earned him less than expected, bought some goods at Fort Yukon and, with his two good teams of dogs, brought them to Circle City and Dawson where, as we have seen, there was an urgent need for all kinds of produce. The journey was tough, but once he had reached his destination, he had no difficulty in earning a good sum.[1] Heartened by this, he repeated the trip five times and made a good profit each and every time that he did it. When spring came, the food shortage was resolved at Dawson, but there was still a great demand for equipment, tools, wooden planks, warm clothing, medicines, and anything else that might be needed in the vast deserted land.

People were crying out for all and sundry in the emerging Alaskan towns, and it was also thanks to the contribution of enterprising and experienced men like Jack Costa that well arranged living areas sprang up, similar to those seen during the gold rush in much more favorable environs, and that people managed to survive in this harsh and empty world, straddling the border between North Western Canada and the Artic Pole Circle, where night ran from November/December to January/February, where in summer the sun never set, where the mosquitoes brought malaria and the temperatures often dropped under 50 degrees below zero so that people could freeze without realizing it, and the reflected sun could burn unprotected eyes.

The first few cabins and houses were also built at Dawson, where saloons, hotels, theaters and brothels were to spring up. Wells were dug deep with a view to ensuring that the typhus epidemic, which had affected the area, would not come back. Doctors arrived on the scene and hospitals were opened, newspapers were founded, a postal service was organized, banks were established and a team of firemen was set up to deal with the flames that continuously broke out because of the fires that were lit inside tents and houses for protection against the cold.

## NOTE

1 *Jack Costa, prev. cit.*

# 20
# A Big Find

Jack Costa, the sourdough and survivor of years of exploration and adventure in Alaska, lived within this new framework, and had turned into a man very unlike the young peasant from Pederobba who, through sheer hard experience, learned to completely master new skills and grasp the necessary know-how. He now knew perfectly well how to steer his dogs, construct a vessel, trade, dig through ice, and find his way confidently in immense stretches of land while defending himself against the cold and wild animals, and recognize the signs of nature and the seasons. He now belonged to a completely different world, and was part of an extraordinary community of men who had come here from all corners of the Earth to follow a dream that each of them had in common; they were competitive and shared the frontier spirit, but were ultimately united in the face of an environment that, with its harsh conditions, was a challenge to them all.

While Jack carried on with his activities, alternating between searching for gold and trading, his brother Frank devoted himself more continuously to digging. There was a period of six months during which Jack knew nothing about his brother. When he had almost lost hope, he found him by leaving signs on the ice;[1] then they both carried on with the search, together or separately trying new areas and calling out to each other to see if either of them had found something good. During the more favorable seasons, journeying was done mainly along the rivers, in simple canoes or boats that were assembled with wooden planks, and then sealed with pitch, with the constant risk of falling into a whirling waterfall or crashing against one of the remaining ice blocks.

When the sourdough was flanked by these large floating blocks of ice, the safest thing to do was to jump on top of them, drag the load and the boat carrying it onto the ice, and let the current carry the block of ice until it settled on the riverbank. He kept a clear berth of the roughest parts of the river and made maneuvers fast so he could reach the nearest shore to save himself from being engulfed by the whirlpools that came just before the rapids. Then the goods and the boat had to be painstakingly transported overland, several trips being made before the waters were calm and navigable again. Another nuisance in the summer months was having to wear a head-net and gloves — right up to the elbows — to avoid being bitten to death by mosquitos.[2]

Throughout the year it was bitterly cold. Even during the warmest summer nights the temperature dropped many degrees below zero and, in these circumstances, once again the dogs turned out to be precious since, they gently huddled close to their master, their thick coats keeping him warm. Warm, yes, but not too hot, because any excess sweat could freeze on the skin and, if it did so, a man would never wake up again. This they had learned from the Indians: work as slowly as possible to avoid sweating through overexertion — otherwise change into dry clothing straight away. The diet consisted mainly of tinned food, dried meat, flat bread and beans, and months could pass without there being another single living soul in sight.

Jack Costa was often at risk, and sometimes had to face life-threatening hazards. Once, when wielding an axe, he seriously wounded his leg.[3] Completely alone, he knew he had to treat the gash by himself, since risked getting gangrene; luckily he was strong and healthy, and recovered without suffering any complications. He then resumed his quest, defending himself at night from wolves by lighting a ring of camp fires around his small bivouac, waking up every two hours to keep them alight, and throwing a few glowing cinders into the woods to show that he was alive, awake and ready to sell his skin — only at a heavy cost. Once the wolves managed to eat one of his dogs[4] but shooting it would have proven useless because as soon as a wolf was shot, others, even more aggressive and determined, would show up.

Like his brother, in late winter 1898, Frank Costa took a serious risk while looking for gold along Seventymile River. Spring was on its way, and he had almost run out of supplies. He had to go down as soon as possible to the Yukon River following the large and still icy course of water because the plain was covered with snow that was too soft and deep for him to cross on foot, and the smaller streams were bursting with water and a dangerous buildup of snow and ice. After a few days' traveling, Frank saw that the ice in the river had started to break up, and decided to use the wooden recipients that he had used during the autumn to sieve the earth: after fastening them together, he made them waterproof, thus creating a makeshift boat in which he could sail along the current. Where the ice formed a barrier so hard and compact that it stopped the raft, Frank held firm on the bank and waited patiently for an opening to appear. But, about halfway through the journey, the waters were so rough that they hurled his vessel against a rock, and then threw it onto the rocky river bank, overturning it. Frank fell into the freezing cold water and was carried to the valley by the whirling current. He was wearing rabbit-skin boots and a heavy waterproof coat that he had to take off immediately in order to swim, with great effort, to the shore. Luck had it that, by running fast along the shore, he soon managed to get his waterproof back, and then he found enough dry wood to light a large fire, by which he could warm and dry up. He then resumed the exhausting journey until he reached the longed for Yukon river. Here he found a safer ship that took him without further hitches to the town of Circle City, the right place for obtaining the necessary equipment to resume his search.[5]

The winter of 1898 was punctuated with various finds throughout Alaska including one of the most significant, a deposit 60 centimeters thick and 60 meters long, which was discovered in the area around Nome, thanks to the Swede with whom Giovanni had shared his first journey inland.[6] That year the efforts and misadventures of the Dalla Costa brothers were, for the first time, adequately rewarded. Jack discovered the longed for gold on Preacher River. It was a large deposit; although not substantial enough to share with his brother, it was big enough to make up the money that, together with the savings from the many small finds and trading activities, would enable him to close his Alaskan venture and go back home, to Italy, to resume his life and put an end to a desire, that long been growing: to have a family of his own and enjoy a peaceful future.

And this was to be. Jack Costa sold his things, freed himself of his claims, drew out the deposits of gold that he had gradually made in the local banks, said goodbye to his brother and friends and, gliding first along the river and then sliding on the waxed cloth used as a sleigh, down by the steep icy and snowy slopes that went down toward the coast, he returned to Skagway. He embarked, and then reached his old house in the mining area of the state of Washington, where he took a break from his long journey back home.

But here something happened. Something bad, very bad. His son Francesco was to recount many years later that his father Jack told him that his friends, whether true or false, sucked far more blood from him than the Alaskan mosquitoes had ever managed to do.[7] But Jack was always very guarded when he talked about the events of those times; perhaps someone cajoled him into making bad investments, or perhaps others borrowed from him without being able to pay the money back; maybe he was swindled or robbed — we will never know — but we do know that in an incredibly short space of time Jack Costa saw his fortune disappear and found that he had bounced back to square one, and no longer had any gold. He was almost overwhelmed by a heart-breaking dilemma: should he start his homeward journey having survived an adventure that had left him almost broke, or should he go back north, and be more careful this time, starting again from scratch. Jack was thirty-one then and, notwithstanding his misfortune, was still bursting with vitality. He decided to give it all another go.

Alone he overcame the terrifying upward slope and the long journey toward Klondike and, soon afterwards, in fall 1899, was on his return, back on the icy stretches whose traps he knew so well, each and every one of them. Here, as well as having to swallow the bitter morsel that life had treacherously reserved for him, he had to eat another piece of humble pie by putting up with the humorous, insistent and inevitable quips from his brother and companions. They certainly made a field day of it when they saw him returning, crestfallen and furious, just a few months after seeing him leave happily and triumphantly for home.[8]

However, soon Jack Costa set things right. Already in the winter of 1899, after a few deals that turned out trumps, he managed to amass enough money to start the search again, with renewed vigor. In the spring of 1900 he left with his brother Frank and Felix Pedro, the man from Modena who continued to be a good friend and collaborator, towards the source of the Cariboo River. With Frank he went back to Circle City in August after leaving Pedro all the food they could do without, and dug on the Faith, Hope and Charity Creeks, but found nothing. So Jack set himself up again as a business man and traded in the outlying areas of Dawson. In the winter of 1901, he passed the confines of the United States part of Alaska and started digging in a patch that he had already checked with little success in 1896; it was on the Miller River, not far from the Tanana River, an important tributary of the Yukon.[9]

## Notes

1 Recollections of Giovanni (Gianni) Dalla Costa son of Francesco, the first born son of Jack Costa. We must bear in mind however that, as remembered by Franco Bonfiglioli, son of Rina, the firstborn daughter of Frank Costa, in those times people found each other through messages left on trees.

2 These extreme environmental conditions were underlined by Lino Pellegrini in the above cited articles.

3 L. Pellegrini mentions this episode in "Un trevigiano 'aprì' la corsa all'oro nel favoloso Yukon," *prev. cit.*, and in the other above mentioned articles. The incident is confirmed in the recollections of the descendants.

4 Recollections of Giovanni (Gianni) Dalla Costa.

5 *Frank Costa, prev. cit.*

6 Lino Pellegrini. "Da solo nell'inferno buio," *prev. cit.* The episode is confirmed in the recollections of Maria Ghidelli, who specifies that this vein was 60 cm wide and 40 cm deep. Jack Costa rested on it during his first arrival without ever imagining he was sleeping on a bed of gold.

7 L. Pellegrini, "After years of adventurous searching, the brothers Giovanni and Francesco Dalla Costa discovered a large gold vein", in *Il Globo,* an Australian newspaper, year XVII, no. 2. January 3, 1975, p. 3.

8 Recollections of Giovanna Franca Dalla Costa garnered from her father, Francesco, Jack Costa's son.

9 *Jack Costa, prev. cit.*

# 21
# Turning Point

The winter was particularly harsh that year. Jack Costa, in constant contact with his brother Frank, and Felix Pedro  despite the biting cold, was sieving the earth in search of the precious metal, along the course of the Saleha, the Beaver and at the sources of the Chena. "He put up with freezing cold fingers, his shoulders and his feet hurt because they were frozen… even the dogs had frozen paws, although they were better suited to these extraordinary conditions."

Then Jack had an experience that he was to remember as the most difficult of all that he had endured in those days in the ice. While making his usual trip to stock up again on supplies, a journey that he made without a tent or a stove, he was suddenly overcome by the cold; it was so intense, at around 55 degrees below zero, that in one night it froze six of his dogs to death. Jack, who realized his situation was extremely precarious, was in the icy desert, but had to face the return journey. Half frozen and overcome by fatigue, he pressed on until he eventually managed to reach his encampment safe and sound.

Even then Jack had the strength to go back to the spot that he was searching, and to carry on digging. After reaching a depth of 15 meters with disappointing results, he decided he would to try a little further ahead the following winter. He returned to Circle City, where he stayed until late summer 1902, when news reached him that he was now needed by Felice Pedroni, who had found gold. He lost no time in reaching Felice, and there he found that things were far better than he had expected.

Not only had Felix Pedro often suffered sheer hardship together with the Dalla Costa Brothers, but he had also shared his secrets - there was one in particular that he loved to repeat by the fireside before falling asleep. It was the story of his Lost Creek, a river where he had sieved using the flat pan system and had made extraordinary finds. But then, since food was scarce, he moved away from the flowing water to reach the provisions store. When he went back, he was unable to find the right road for the stream in the thick of woods, hills, valleys, rivulets and mountains so typical of the area surrounding River Chena. This was in 1898, and for the four years that had passed since then, Felix Pedro had done nothing but think about his Lost Creek and try, time and time again, to find it. In spring 1902, almost at the end of his tether, he and his companion for the season, Tom Gilmore, had only a hundred dollars to call their own. They tried to get equipment, victuals and credit at the nearest store. Chance had it that the store owner, Barnette, who was also captain of the steamship Isabelle, was away on one of the long trips he used to make along the river searching for goods to stock up his store, which he had left under the management of his wife's brother, Frank Cleary. And this was how a twenty-five year old youth, a civil engineering student at the University of Montana, whom Captain Barnette had firmly told to give credit to none, did things off his own bat and unwittingly made a fortune for the entire community. He took the hundred dollars given him, and turned a blind eye to the outstanding amount. This meant that Felix Pedro and Tom Gilmore could continue to search and dig in the same spot that Jack Costa and his brother Frank had already searched and dug in.

After the encouraging find of traces of gold ore on July 2, the diggers went back to the store to ask Cleary for some food and medicines and again were given goods on credit. Then Gilmore went off to Circle City to sort out a few things, and Pedro stayed on his own to dig — and kill mosquitos. On July 28, Cleary was amazed to see Felix appear at the door again: this time he would have refused him any further credit. But Pedro had something other than supplies on his mind. He was excited, very excited. He took Cleary to one side and told him that he had found it: yes, a few days back, on July 22, he had found gold. But he was being cautious. He wanted to be very reserved and intended to tell only his true friends the news in order to prevent the rush of opportunists and profiteers that he had already witnessed when he thought he had found gold at Lost Creek and had spilt the beans, telling people that he had at long last hit the right spot. On that occasion he was almost pleased when things turned out otherwise because, drawn there by the interesting news, within a few days this flock of people had gathered together, ready to pounce on the gold, so much so that he felt desolate and confused. So this time he bit his tongue and told a few trusted friends.

And this is how it was. Among the first to arrive were Jack and Frank Costa, who helped him verify the richness of the deposit deeper down.

Things seemed really good, with the pans containing material worth well more than fifteen cents, and so they soon applied for claims.

How did the claims system work? It was simple. The first find on a river was numbered zero, the starting point. The other claims, were numbered progressively starting with number one both above and below on the creek. The claim, a rectangle measuring forty meters by fifty, was marked off by a fence with four stakes in the corners and a poster bearing the name of the claimant. So Felix Pedro, apart from christening his river "Pedro Creek" and the adjacent hill "Pedro Dome," managed in the space of thirty days after the find not only to obtain zero, the "Discovery on Pedro," but also the license to search another four streams that flowed around the Pedro Dome highland: the Cleary, the Gold, the Twin and the Gilmore.

In turn, Jack Costa obtained number two above on Pedro, number eight below on Cleary, number seven above on Gilmore, number sixteen below and number five to the far left on Fairbanks, number one above on Treasure and number nine below on the Dome.

Frank then got the claim for positions number three above on Pedro, number one above on Bear, number seven below on Cleary, numbers fifteen and eight below on Fairbanks, one below on Kokomo, six above on Gilmore, nine above on Fox and eight below on Dome, Discovery and Treasure.

When the inner group of friends was happy with their claims, the news of the find was let to skid along the ice, above all by Captain Barnette, who, more interested in selling his goods than in digging for gold, was delighted to see hoards of prospectors flowing into the area. Barnette returned from his journey along the river on September 8, a month and a half after the find, but he soon made up for lost time. Above all, the provisions that he had brought along in his boat could be used for throwing the largest possible celebration lunch, his guests feasting on Indian and western foods and drinks: the menu listed a dish based on *hiyu* and *muck-a-muck,* to be accompanied by a blend of spirits christened for the occasion, Chena cocktail and then tawtilla consommé, olives, chicken mayonnaise, ostrich paté, sauterne, wine jelly, cream sauce, "seeker" style roast moose, potato and green pea puree. This was rounded off with ice-cream, cake, walnuts, raisins, cheese, coffee, cigars and canned yuma. It was an unforgettable party, but time was marching on, and there was no more left to waste.

Two days after the ship's arrival, in fact, a meeting was held between the gold seekers in Felix Pedro's tent, and during it Barnette not only managed to get himself elected as provisional administrator of the district in order to keep the contracts register, even if most of the miners continued anyway to deposit their contracts at the Circle City Registrar's Office, but he also managed, with the consent of those present, to christen Fairbanks in honor of the Senator about to become vice-president of the United States, the city that without doubt was about to rise around his own wooden warehouses on the river Chena.

In a short space of time hundreds and then thousands of men turned up in the valley of Tanana. This stampede could not have been foreseen, but behind it was the hand of Captain Barnette, who, apart from sending enthusiastic letters to acquaintances and relatives, deemed that the hour had come to put the best agent he had into action: Jujiro Wada, who was known by all and sundry as "Wada the Jap."

In 1890 Wada, an ex-marathon runner, had boarded a whaling ship in Japan to reach Alaska where, after years of adventures and trading, he became cook on board the Isabelle, Captain Barnette's ship. Wada's ability to cover long distances without provisions was legendary and he had become part of the Alaskan myth, thanks to his journey from Hershel Island at Icy Cape, when he covered several hundreds of kilometers without food, and in mid winter. He was a great character "Wada, the Jap," very well known and highly respected; he was just the man to make Captain Barnette's plans bear fruit. Three days after Christmas 1902, Wada was sent by the captain to Dawson City, in Canada, with a load of letters and skins. Nineteen days later, anticipated by so many stories, endlessly blown up out of all proportion by word of mouth, and tales of the Bonanza of Tanana that had skated along the ice, Wada's arrival in Dawson got more people moving than any disembarkation of the King of England could ever have done, in all its pomp and ceremony.

With his good nose for sensationalism, Wada urged people to leave straight away. He did the work captain Barnette entrusted him with so well that the local newspaper, the Yukon Sun issue of January 27, 1903 came out with the headline "Great find in Tanana." Complete with a map and pictures of Wada dressed in a huge fur-coat, the article went on to tell its readership that the Tanana findings were worth as much as those made around Dawson, and that exploratory diggings had yielded a series of fantastic successes. The lines that followed extolled the trustworthy and conscientious personage, Wada himself, saying how he was well known throughout the north and how he was, above all, a sincere and honest man. Wada himself rounded off the piece by declaring "In seven months' time all the embarkations downstream on Tanana will go back upstream again because there's another Klondike in that village."

It was a freezing cold winter, the temperature dropped during those days to 47 degrees below zero and, not far away, in the town of Fortymile, to minus 59 degrees. Yet Wada's efforts yielded an outcome that went way beyond the wildest dreams of Captain Barnette, who soon found he was catering for not only more than one thousand men, whose pressing demand for goods exceeded his current or potential future supply, but he was also contending with their righteous indignation at being drawn into an adventure completely at odds with the picture given by Wada. The Japanese man, on his part, narrowly escaped the gallows erected for him by men from Dawson, whose mood changed to blind fury when they reached Tanana.

## NOTES

1 Relatively recently, the collaboration between Frank Costa and Felice Pedroni was recalled in an issue of the *Alaska Sportsman*, September 1961.

2 Recollections of Maria Ghidelli.

3 *Jack Costa, prev. cit.*

4 James Wickersham, *Old Yukon. Tales-Trails-Trials*, Washington Law Book, Washington, D.C., 1938, p. 210.

5 Alex Hanot, *op. cit.*

6 Whereas in the article "Jack Costa," *prev. cit.* and in Terrence Cole, *op. cit.*, chapter IV, claim number two is cited as being situated above on Pedro, the caption on the photograph of the mine (a postcard of the spot see plate *IX*) in the hands of Jack Costa's descendents states: "Costa's claim n. 2 below on Pedro." We established that the caption on the postcard was erroneous since stake no. 2 below belonged to Tom Gilmore.

7 *Jack Costa, prev. cit.*

8 *Frank Costa, prev. cit.*

9 L. Pellegrini, "Il modenese che vinse la corsa dell'oro" [The man from Modena who won the gold rush], in *Domenica del Corriere*, no. 38, September 19, 1965, p.33.

10 Terrence Cole, *E. T. Barnette. The Strange Story of the Man Who Founded Fairbanks,* Alaska Northwest Pub. Co., Anchorage, Alaska, 1981. Although the text does not specifically cite the names of the participants at the meeting, it is reasonable to assume that the Dalla Costa brothers would have taken part in it.

11 Terrence Cole, *op. cit.*

12 Terrence Cole, *op. cit.*

# 22
# Victory

What did the mass of men who had stampeded from Dawson to the Tanana fields find? Why were they so angry that they were on the point of preparing a noose for Wada? There were two main reasons: above all, Barnette's stores contained enough to meet the needs of tens of men, perhaps a few hundred, but not the needs of the thousand or more who reached this place, panting along for almost three hundred miles, encouraged by the Japanese ex-marathon runner. But, all in all, this problem was peanuts to the prospectors, who, hardened to deprivation and the cold, were ready to dig barehanded if they as much as smelt the precious metal.

They were infuriated above all by the fact that the gold wasn't there. Or rather, the prospects were positive and the sieves yielded more than ten cents' worth a time, but the hefty deposit, the final proof, had not yet been found. They just had to suffer and toil, in an endlessly freezing cold winter that bit their faces, hands and feet, with little or nothing to eat and hardly any equipment. Many gave up, and others stayed on. But in spite of it all, the work continued, particularly in location number two above on Pedro, where the Costa brothers, Jack and Frank, constantly sieving pans worth more than fifteen cents, felt that something good, something really good, lay down there under the ground.

Around Christmas time Captain Barnette had written to a friend of his in Seattle, Jim Morrison, in terms that were perhaps somewhat over-emphatic but that were anyhow a sign of how promising the situation was: "Currently we have the best places in the whole of America. Pedro Creek, the course of water where the most work is done, continues to be increasingly profitable. I have just returned from location two above on Pedro. The owners wanted two thousand dollars' worth of equipment, so I have gone to see if the spot where they are digging could provide good guarantees. There were seven shafts, and though they had not yet reached the rock at the bottom, they have already collected gold that will be worth more than fifteen cents per sieve. Wada says twenty, but anyway it's much better than I expected. It is the most lovely thing I have ever seen."

Then, to reassure his friend, who was probably well aware of his overoptimistic nature, he added: "What I have told you is true and the first to get here will be the one to get the best place."[1] Maybe Jim Morrison was one of the thousand or so men who were undecided only as to which of the two to kill first: the captain or his trusty ex marathon runner, Wada. This, of course, we will never know for sure, but we do know that Jack and Frank Costa were investing heavily on stake number two above on Pedro.

At first they had not believed in this stake a jot more than they had in any of the others that they had scattered around, and that were fruitful, but this one had something special about it. So down they went, full throttle!

But what did it mean to work in that harsh winter? It was a titanic enterprise. First they had dug four shafts and then another three to make a total of seven. They were circular holes with a diameter of about 85 centimeters. The method was always the same. They had to thaw the land with the only material that was plentiful: wood, set alight with paper or rags steeped in kerosene at the bottom of the shafts. When the flame had done its work of thawing the land for a few centimeters, a man could drop down into the hole and, using a pickaxe and shovel, remove the gravel down there. Things went well when the men managed to go down about twenty centimeters a day, but work proceeded more and more slowly the further down they got because, once the flame had gone out, the smoke lingered longer in the deeper holes making it impossible to go down any further to remove the thawed earth.

To have a good picture of the scene in those times and the atmosphere in those places, we can read the descriptions of the excavations made on the Bonanza and Hunker creeks, in Klondike, issued only six years before by the mining engineer T.A. Rickard:

> If you had gone up the narrow valleys of Bonanza and Hunker during the long twilight of the arctic winter ten years ago you would have seen a picture worthy of Gustave Dorè. There was no noise, for there was no machinery; there were no whistles to announce the noon hour or the evening rest; there was no drilling in hard rock nor cheerful hammering. A weird silence brooded

> over the waste of snow. The gloom was thickened by a pall of smoke escaping from holes in the ground, whence an occasional figure issued. Not many men were visible, for they were below in their warren, like digging rabbits. At the top of a shaft, here and there, a weary gnome might be espied turning a windlass and emptying buckets loaded with dirt that came from a small pit beneath. The flare of red fires parting the twilight marked the beginning of the work of shaft-sinking. The snow, the moss, and the fog muffled every foot-fall, deadened every sound. It looked like hell, but it was freezing.

And so, day after day, for that whole long winter, Jack and Frank dug their seven shafts in concession number two above on Pedro Creek. They went down meter after meter, to a depth of ten, fifteen, so much so that Jocko, Jack's dog, was anxious and in distress until his master came back to the surface.

It continued like this until the morning of April 9, 1903, when Jack made the discovery he had sought for more than ten years.

The event was told through the eyes of an exceptional witness, Judge Wickersham, author of the book Old Yukon, who happened to see this extraordinary happening on his way from Rampart on thc Yukon River to the new town of Fairbanks, where he was opening a court of justice:

> Rough-locking our sleds with dog chains we plunged down the ridge between twin creeks, and through the forest, following the blazed trees until we reached a little clearing at Costa's cabin on Pedro Creek. As we stopped our husky team before his cabin, big Jack Costa came up the ladder out of a prospecting shaft near the doorway. His chubby round face was distorted and wrinkled with excitement and deep feeling, and, without knowing or caring who the stranger was standing by his shaft, he cried out, his rough bull voice vibrating with excessive joy:
>
> "Oh, by Godda, I gotta de gold!"
>
> He had just struck pay in his shaft, the first to be found on this part of the creek, and was fairly overcome with his sudden fortune.

Luck? We know only too well that luck goes only so far in this story. It was, above all, the courage and determination of Giovanni Dalla Costa that underpinned his achievement. But further on in his account, Judge Wickersham himself underlined Jack's qualities, saying he:

> . . . had dug holes on Forty Mile, Birch, Faith, Hope, Charity, and a hundred unnamed creeks in the Tanana hills, but just now, after years of labor and failure, Fortune smiled! Visions of Italy, the old home and the old mother, the girl he left behind when he came to America, a vineyard and a winepress, wife and children — big Jack cried as he babbled to us of these things and the gold in the pit at his door.

## NOTES

[1] Terrence Cole, *op. cit.*
[2] Terrence Cole, *op. cit.,* p. 41.
[3] *Jack Costa, prev. cit.,* p. 19.
[4] James Wickersham, *op. cit.*, pp. 180–181.
[5] James Wickersham, *op cit.*, p. 181.

# 23
# Bucketfuls of Gold and Homeward Bound

So there really was gold in claim number two above on Pedro Creek and the Dalla Costa brothers could at long last have a clearer idea of their destiny. Soon they set up a mining company to all effects, with workers and just one sluice at first, shortly to be followed by a second one. The mining was opened in all directions, with no less commitment and effort, but at last with the knowledge that the work done would bear fruit.

This discovery, which attracted many other seekers from all over Alaska, was the main lever for the development of Fairbanks, which was now destined to become a city true and proper, second only to Anchorage. Three years of intense activity were to follow for Jack and Frank, during which time the brothers' estate grew constantly as did their friendship with Felix Pedro, who, while telling stories during the long arctic nights, let out more and more information about his past.

It was in this way that, before a warm fire, Jack and Frank learnt that Felice Pedroni was born in 1858 in Trignano, in the municipality of Fanano (province of Modena). The son of a poor peasant family,[1] and the youngest of six brothers, Felice was orphaned at the age of six and, when only a child, had started to work as a shepherd in Veneto and Tuscany and had done a bit of hoeing in order to help his household out. But this miserable life was not enough for him, and he was courageous and enterprising enough to think big: in 1881, at the age of twenty three, he followed the example of his other brothers who, one by one, had gone on searches that took them far from their homeland, and had left Italy for the United States. He had drifted in Illinois, Oklahoma, Colorado, Utah, Oregon and the state of Washington, doing odd jobs, working as a manual laborer on railroad constructions, and as a miner in various places until he reached Carbonado, where he met the Costa brothers.

He told them not only about the joy and bitterness of his Lost Creek, but also of the time when a snow storm had killed two of his dogs and he, short of supplies, hungry and alone, had eaten one of them to keep himself alive. And of when, thanks to his good aim, he managed to kill a big bear, and then used its skin and paws to make himself two warm moccasin boots that saved him from frost bite. And of another time when he encountered a group of geographers who were desperately trying to draw a map of those areas. And it was he who, uneducated, helped the technicians to set the drawings right. For this help he was given an extremely useful present: a pair of leather boots with rubber soles. And yet again, of when he killed a moose and found a gold nugget wedged between its toe nails. The story had it that, by following the backward tracks of that animal, he had hit right on Pedro Creek, just where they were digging and earning bucketfuls of money.

Those tales were told by Felix Pedro, and the Costa brothers responded tit for tat by telling him of their ventures and experiences. But Jack's stories were hinged increasingly often on his life back home in Costa, on Pederobba, the hill of Monfenera and the river Piave, places that he yearned to see again. And so, day by day, nugget after nugget, the nostalgia for his origins grew greater and stronger until it became uncontainable. His momentous decision was made in 1905.

After saying goodbye to his brother and Felix Pedro, with a large quantity of gold deposited in a United States bank — this time he did not fall into the clutches of swindlers lying in wait for miners returning from the gold fields — and, after a speedy and smooth journey, he reached Pederobba after an absence of almost nineteen years.

## Notes

1 Lino Pellegrini, "Gli italiani e la febbre dell'oro," *prev. cit.*, "Il modenese che vinse la corsa dell'oro," *prev. cit.* and "Un orso lo salvò dal gelo" [A Bear Saved Him From the Ice], in *Oggi*, year XXX, no. 51, December 18, 1974, pp. 75 ff. In Alex Hanot, *op. cit.,* it is stated that Felice Pedroni was born in 1860.

2 This incident was described by Lino Pellegrini in *Gli italiani e la febbre dell'oro, prev. cit.*, p. 110.

# 24
# A Fabulous Honeymoon

Of course Jack was unlikely to have read the Odyssey, but he arrived back home incognito, just like Ulysses, to see how his fellow countrymen might behave towards a stranger in difficulty. On his return he pretended that he was poverty stricken and in need of a bed for the night, and this is how Jack Costa spent his first night back home: the man who had made a fortune in Alaska, and had returned to his birth village to start new life, spent night number one on a hay stack in a cowshed.

The following day, however, people started recognizing him. One of the first to work out this stranger's true identity was Rocco Rostolis, Giovanni Dalla Costa's godfather, a foundling, but loved by all because, although he may not have been, as word had it in the village, the love child of a noble who had consorted with a commoner, he had a strong sense of elegance and gentle ways, and he showed that he had an admirably noble spirit on all occasions.

Seeing that he had been found out, Giovanni decided to show himself for what he really was, a well to do moneyed gentleman bursting at the seams with gold nuggets. There were no banks at Pederobba in those days, and so he decided to deposit many of his riches at the Società Operaia di Mutuo Soccorso [Workers' Mutual Aid Society] of Valdobbiadene, beyond the river Piave, a decision that was to profoundly influence his future life. It was a legendary deposit for those times, ninety thousand lire in gold coins. He also bought a large, solid noble house, four centuries old, with walls that were a meter and thirty centimeters thick, which until recently had been set up as a barracks for the carabineers and a prison, in which he had extensive restructuring work done. Together with this house he bought the entire complex of houses that went along one side of the main street, right up to the home of Doctor Calvi, the medical practitioner of the village. On the other side of the street he bought another beautiful and spacious house adjacent to the

Church square and then fields, land to cultivate, woodland and a coach and horses.

He used to hand out nuggets for services given, giving two to a worker who put a small safe into his home, never hesitating to give work and recompense to his fellow countrymen whom he found to be more or less as poor as he had always remembered them. But his main concern was to marry, and the chance came quite soon, following a second lucky meeting in Padova with his Godfather, Rocco Rostolis, who told him how the work of his family was faring.

Rostolis and his family had an artisan activity, working wool at full cycle. Sheep breeders, they sheared wool, washed it, sent it off to be spun, re-washed it, and then spread it out on lunettes, trestle-like structures, to dry. Then they took care of making twine and weaving cloth using pedal looms that they had placed in their old granary. Raimondo, one of the children, had opened a tailor's shop and this brought the circle to a perfect close: from sheep to tailored suits.

After telling him about these innovations, Rocco invited Giovanni to pay a visit to his home and see for himself how things worked there. And, a few days later, Giovanni knocked on the door of the Rostolis family home to see Rocco again, and to meet his wife, Regina, and all the children, with their habitual R at the beginning of their names: Renato, Raimondo, Raffaele, Roberto, Riccardo, Romano and Rosa. It was a memorable occasion. As soon as he crossed the threshold, Jack Costa started giving the Rostolis family members pieces of gold, and gold necklaces that he had in his pocket. One nugget also went to Doctor Calvi, invited to the gathering.

But Giovanni's eye alighted on his own nugget, Rosa Rostolis, called Rosina, daughter of Rocco, graceful, reserved, thirty-two years of age; her last seven years had been enriched by her copious correspondence with a Brigadiere from the carabineers, a certain Berto, who was destined to be the father of the writer, Giuseppe Berto. But the persistent courtship of Giovanni, his generosity toward the family, and the desire of the Rostolis family to see their daughter married to a real gentleman, held sway over the epistolary love. They persuaded Rosa to accept the marriage proposal notwithstanding adamant opposition from Berto who, on discovering what was happening, put a stop to the correspondence with a furious letter that ended with a sentence to this effect: "May the Earth gape open and swallow you up whole for this."

But now, however, Rosa's future had been decided upon, and the wedding was celebrated in peace by all on February 1, 1906. It was a great ceremony, and the entire village participated in the festivities, talking about the event for a long time to follow, but the really singular and extraordinary event was yet to come. In mid April the newly weds left on their honeymoon. Where were they bound? Alaska of course, and this time not as emigrants, but as gentlefolk, in ease and luxury. The outward journey lasted three months, first Paris and then the port of Le Havre, where they boarded the Le Touraine, which took them to New York, where they disembarked on May 13. Then the journey was resumed by train, a forty-day trip across the United States that Rosina remembered particularly well: she was greatly impressed by the fact that at mealtimes the convoy stopped so that the kitchen ware and tables could be set down on steady ground for food to be prepared, and meals served to the passengers.

The newly weds visited Boston, Philadelphia, Washington, Chicago and finally San Francisco. From here first in an ocean ship and then in one of the romantic steamers, hundreds of which were by then sailing along the Yukon, they reached Frank Costa who, on hearing of the arrival had made arrangements to receive them as well as possible. It was an exciting journey, during which Giovanni and Rosina conceived their first child, a son, whom they were to christen Francesco, after his uncle.

They also met Felix Pedro who, to Jack's great surprise, said that he had also been back to Italy in recent months, to Trignano, but that things had not turned out so well for him there. Forty-seven years old, having been away for twenty four years, he failed to trace his old friends, and nor had he found love. He had tried to court Egle Zanetti — she was a school teacher and that fascinated him because he was only just about able to sign his name — but in spite of his serious and repeated proposals of marriage, he was rebuffed. The people of Trignano themselves had not really believed his stories of gold digging or his Alaskan wealth. So, a few months later, bitterly disappointed, he returned to Alaska.

Rosa stayed in Fairbanks for about two and a half months and, during this time, she had the chance to realize that the riches of her husband had come from the intense cold, even though it was the end of June, from the mosquitos with their incessant biting and from the cabins, which were far less comfortable than the one her kind brother in law had built for them, more spacious than the others and even equipped with a really warm bed made of bear and wolf skins. She saw the mud and the shafts and the strenuous digging work, but she proved worthy of Jack, taking everything in her stride, and adapting well to the situation.

It was probably during these magic moments that Giovanni made a present to his spouse that was destined to stay famous for generations to come: he told her to choose from a pile weighing about two hundred kilos stacked in the corner of the cabin as many nuggets as she wanted, all of the same size, to make a necklace. The end result was a chain about two meters long that Rosa wore wrapped around her neck three times over, though she did take the precaution of wearing a special bone collar to bear the weight.

Jack and Frank both agreed that it was time to end their Alaskan adventure. They were at the height of their success. Stake number two upstream on Pedro Creek had already earned them 150,000 dollars and could still have yielded at least the same amount again. They decided to sell it to Ed Bartlett, a business man who already owned an hotel and would have exploited the remaining gold deposit. Frank was to remain there until the ongoing business arrangement was settled, and then he was to return to Italy. In late November, the couple said goodbye to Francesco and Felix Pedro and left.

The honeymoon of Giovanni Dalla Costa and Rosa Rostolis was truly golden. An unforgettable period that Rosa only described in part, relating special moments that had impressed her more than others. Above all she was to say that the journey was too luxurious, and that they were always served by black people wearing white gloves. Then she said that on the boat bound for Alaska, there were groups of women getting on board. Her curiosity aroused, she asked Jack who they were and he responded somewhat evasively and so she realized that these women were prostitutes. Finally, she told her listeners about the clay pots full of nuggets that covered the floor of the cabin, under bales of straw, strewn all over the place. Instinctively she counted them, and of an evening she noticed that the number that she had counted in the morning had usually gone down. And then the door was always open, and Jack didn't seem particularly bothered by this. But she knew the value of money, and didn't like this habit of Jack's at all.

## NOTES

[1] These recollections regarding Giovanni's return to Pederobba were garnered by the niece, Mariarosa Dalla Costa during her talks with her aunt Rita, Giovanni's daughter.

2 In 1937, this Society, which had been active since 1880, became the Banca Piva di Valdobbiadene [Piva Bank of Valdobbiadene].

3 Currently the house corresponds to no. 28, Via Roma, Pederobba.

4 Currently the house corresponds to no. 1, Via Roma, Pederobba.

5 For several years Giacomo Bresolin, nicknamed Metto, served Giovanni Dalla Costa as his coach-man and trusted servant. Many of the villagers remembered Metto for the unusual nature of his work in that setting, just as they remember the smart family that passed by inside the coach.

6 Recollections of Giovanna Franca Dalla Costa garnered from her aunt, Rita.

7 Recollections that Rosmary and Remy (Gianfranco) Bresolin garnered on listening to their mother, Bettina, Jack Costa's daughter.

8 *Ellis Island Records* at www.ellisisland.org. Giovanni's second name was erroneously written down as "Corta."

9 Recollections of Valeria Bresolin, called Wally, the daughter of Reginetta, the youngest of Giovanni Dalla Costa's children.

10 Charles W. Adams, *A Cheechako goes to Klondyke*, Epicenter Press, Fairbanks, 2003.

11 Recollections of Valeria Bresolin and her brothers Luciano and Diego. Afterwards, the necklace was divided into four equal parts for Giovanni's children: Francesco, Bettina, Rita and Reginetta.

12 *Fairbanks Daily Times*, August 30, 1906. In the newspapers of those days, news of the Dalla Costa brothers appeared. The *Fairbanks Evening News* of August 30, 1906 reported that "Golden City [the nickname for Fairbanks in the days of the gold rush — Author's note] has been lively this summer. Large amounts of money are now circulating thanks to the efforts of the Costa brothers and the others on the upper slopes of Pedro Creek: each of them are using 10 to 30 workers." The *Fairbanks Daily Times* of August 25, 1906 wrote, on the other hand, that "Frank Costa on the first day came into the city from Pedro for business." Probably business of the type commented on in the *Fairbanks Daily News* of November 23, 1906, which records that on September 1 Frank Costa bought, at a price of 350 dollars, a building that he sold on September 19 for 950 dollars. In the *Fairbanks Daily Times* of August 31 1906, it is mentioned that John Costa with his wife and brother had left for Dawson the night before on a ship named *Monarch*. The *Fairbanks Evening News* of September 12 1906 comments on the happy arrival of this ship at Dawson after a voyage of only twelve days, and publishes the list of passengers, which includes Jack, Rosina and Frank.

13 On the numbers of women prostitutes in the gold rush territories of Alaska, see Morgan Lael, *Good Time Girls of the Alaska Yukon Gold Rush*, Epicenter Press, Fairbanks, Alaska, 1998.

14 Recollections of Rosmary and Remy Bresolin. Stories of these fabulous pans bursting with gold nuggets are also told in the various articles by Lino Pellegrini, cited above.

# 25
# The Tragic Tale of Felix Pedro

After his unhappy experience in Trignano, following Jack and his wife's return back home, and now that his great friend Frank also had made firm plans to return to Italy, Felix Pedro realized he was getting richer and richer but more and more lonely. So, before the end of 1906 he paired up with Mary Doran, an Irish woman who ran a lodging-house in the vicinity of Dawson, on the border between Canada and Alaska. Between late 1908 and early 1910, Felix Pedro had to sell his gold mine, ceding the claim as best he could. This was a real knock on the head for him. It is not known what disease he had, but on August 22 of the same year, eight years after his first gold find in Fairbanks, Felice Pedroni died while hospitalized in this city.

Only in the nineteen-seventies, beset with requests for information by the inhabitants of Fairbanks who wanted to know more about the founder of their city, did the inhabitants of Trignano realize how important the work of their fellow country man had been, and some of them decided that they needed a clearer picture. Two particularly keen investigators looking into the circumstances of Pedroni's death were Amato Cortelloni, a notary from Fanano, and Lino Pellegrini, a Veneto journalist writing about the gold rush, who had already published several articles on Felix Pedro and the brothers, Jack and Frank Costa.

In his breakdown of the event, which appeared in 1974, Pellegrini reported the revealing testimony given by the descendants of Gambaiani, a fellow country man who helped Pedroni in his gold fields. "Pedroni — they said — suffered from no disease whatsoever. Some decades back, after his return from Alaska, Gambaiani stated that he had gone one evening to the hospital where Pedroni had been admitted to say hello to him, but was blocked on the way to Pedroni's bedside by people who, he believed, were Mary Doran's relatives. His pistol, and in those times everybody carried one, was taken from him and he was thrown out of the building. The following day, Pedroni was dead." When Cortelloni questioned Gambaiani about the founder of Fairbanks, the poor man, struck with a paralysis and aphasia, managed only to mumble: "wife… poison."

So suspicion upon suspicion, but slowly the knot began to unravel thanks to Pellegrini who, in spring 1972, through a colleague in Fairbanks and with the help of another man from Modena who emigrated to California, located the grave of Felice Pedroni, in San Francisco, in the Holy Cross cemetery, Dormitory T, Station 12, grave number 456789, area 40. On digging the grave up there was a surprise: Felix Pedro was there alright, inside a colossal outer casket of lead, resting four thousand kilometers away from the place of his death. Inside this there was another casket and again inside this one, as in a Russian matryoshka, lay the coffin, its upper part in crystal, showing the corpse, which was intact, looking like it had been embalmed; Pedro still had a moustache and was wearing a tail coat.

In October 1972, the three caskets reached Genoa to be transported to the cemetery of Fanano where, on the 12th day of the same month, the entire population paid a tribute to the fellow citizen whose merits they had never acknowledged while he was alive. Before the burial, however, Pellegrini related that the notary Cortelloni had an autopsy done on the body and then an x-ray, and the report provided was to solve the mystery: a pointed metal device, probably a hat-pin, was wedged between two cervical vertebrae. According to Pellegrini, Mary Doran had obtained a death certificate stating that the cause of death was a heart complaint. Then, two years later, she had the body taken to a cemetery in San Francisco in an attempt to make sure that there would be no trace left of it. As far as we know, this widow lived the remaining eighteen years of her life as a wealthy woman. Felix Pedro, on his part, still lives on in the city of Fairbanks, not only in the large wooden pillar with his face on it in bas-relief and the caption "Felice Pedroni, the first to discover gold in the area of Fairbanks," but also in the bust situated in the interior of the local university, and, above all, in the celebrations held every anniversary of the foundation of Fairbanks that every year, on July 22, the date of the discovery of gold there. On this day, livened up with commemorations and period-piece costumes, all those in Fairbanks remember the time when Felice Pedroni, alias Felix Pedro, triumphantly rode into the bank on horseback to deposit his saddlebag that was bursting at the seams with gold.

## Notes

1 L. Pellegrini, Un orso lo salvò dal gelo"[A bear saved him from the ice], in *Oggi*, year XXX, no. 51, pp. 75 ff.

2 Giorgio Comaschi and Claudio Busi, now working on reconstructing this story, stressed that various aspects of the affair seemed uncertain and that no concrete proof was available.

# 26
# Peaceful and Industrious Years

Giovanni's life in Pederobba, on the other hand, was sheltered from ugly surprises. Rather, things had sometimes turned out trumps, the best time being when he tracked down his brother Gaspare, whose address he probably got through the contacts that the parish priests had with church people who had crossed the ocean to go help the Veneto colonies in Brazil.

Gaspare lived in Guaporé, where he had gone to meet up with his father Luigi and mother Teresa a few months after their arrival in Uruguay, probably passing through the Port of Rio Grande, Porto Alegre, the Tacquari River, and then by pushing on for another fifty kilometers or so into the forest. The family members had never worked as settlers, but had first participated in the construction of a road in the Rio Grande Do Sul and then ran a "hay house" which, as we know, was a stopover place where travelers could refresh themselves and stock up on hay for their mules — usually they had seven plus one or two spares — used for drawing carts and wagons along the roads that were being opened up bit by bit in the forest. Guaporé was a strategic spot, and there was plenty of work there. Gaspare had married Virginia Gotardo and had three children by her: Aurelio, aged three, Francesco aged two and Rosalia aged one.

Of course this was good news, but there were also some sad tidings. Both parents had died and there was no trace of the sister, Elisabetta, who had completely disappeared. Grateful for hearing about events in his family of origin, of which he had come to know little or nothing in the last fifteen years, Giovanni immediately wanted to share his fortune with his brother, and sent him a sum of money.

But let the protagonists in this story speak for themselves in the words contained in the letter sent from Giovanni and Rosina to Gaspare around the middle of April 1907, in which they describe their circumstances:

> Francesco is likely to return next fall but it is not known exactly when because there is a lot of business to take care of and no-one else can do it on his behalf. We've got some pleasant news for you, dear Gaspare: about a month ago we were given the consoling grace of a little baby boy whom we have named Francesco Luigi Rocco, after the two grandfathers and the brother who, for the time being, is far away from us. But we call him Cesco for short. Nor are we as yet sure about making a trip, as you suggest, to Guaporè since, as you will also understand, the journey would be a bit too difficult for a person with a wife and children and Rosina certainly would not leave me behind alone and nor would I dare distance myself from her since we feel true affection for each other, and are in perfect agreement and enjoy the greatest possible peace of mind together. So much so, that ever since our marriage, which completed its first year on the 1st day of February, we have never parted for a moment, so parting from each other for a long time, as you can imagine, is out of the question. For me it will be difficult, but let's see what Checchi (Francesco) thinks about it when he comes. Anyway dear Gaspare, if things continue to go well, rest assured that you will never be neglected by your brothers and, whenever possible, they will certainly help you. You have again asked for information about brother Giacomo. We never received any news on him, only through the Municipality did we learn that his whereabouts are in Nice (France) and that he wanted his birth certificate because he was getting married.

In November 1907, Francesco Dalla Costa went back to Pederobba. Before four months had elapsed, he had married twenty-seven year old Maria Poleselli from the nearby village of Onigo, and with her, he left again for Alaska.

> Not to return until he has cleared up every single part of the still unsettled business, which will be wound up in two or three years at the most,

as written by Giovanni and Rosina in the letter sent to Gaspare on 9 July 1908, which goes on to say:

> We hope that things go well, and if they do, then rest assured that you will be remembered by your brothers; if, for any reason, you are in need please turn to us and we will do all in our power to help you whenever. We often have news of Checchi. He is well, and so is his wife who, from what he writes to us, also seems willing to start a family.

But Francesco's stay in Alaska was much shorter than this. In November 1909, the letter written by Giovanni and Francesco together with their wives Rosa and Maria, informs Gaspare in Brazil of the fact that:

> Brother Francesco returned from Alaska more than two months ago together with his wife and a baby of about one year old, Rina.

The letter continues, saying that:

> They are in good health but the affairs in Alaska this time unfortunately could not have gone worse, so much so that rather than making the profit we expected, this journey incurred quite heavy expense, which cannot be recuperated since we have established that there will be no return, and we will content ourselves with what we have and live peacefully and economically as a family rather than embark on new hard journeys. So let us leave aside the desire to make a fortune, which was always strong in us before we settled here, and let us rather lean on the proverb, he who is contented is rich.

## Notes

1 Letter from Giovanni and Rosina Dalla Costa to the brother and brother-in-law Gaspare. Pederobba April 20, 1907. Letters sent from the family members

of Pederobba to Gaspare in Brazil were available thanks to the precious collaboration of Marly Dalla Costa, daughter of Osorio Giacomo son of Gaspare, and resident at Guaporè. On the other hand, the letters sent to Pederobba from Guaporè were destroyed during the First World War.

2 The couple's arrival at New York in the ship *La Provence,* on which they sailed from the port of Le Havre was recorded on March 7, 1908 in the Ellis Island Records (www.ellis-island.org).

3 Letter sent from Pederobba.

4 According to the recollections of Francesco Bonfiglioli and his wife Emanuela (Emy), the name of the first child was registered by an Alaskan official who anglicized it into Rena, the name being officially transcribed this way in Italy. Francesco and Emanuela also told us that this daughter was particularly close to her father and took care of the management of his activities in Italy, thus trying to make amends for the fact that Francesco never had the son that he had longed for.

5 Letter sent from Pederobba.

# 27
# The Families Thrive

And indeed the Dalla Costa brothers kept their word. Francesco had a strong entrepreneurial spirit and was, as ever, determined to make good fruit of the technical know-how he acquired in the three years of military service in the Genio section, and the ten or more years he had spent in Alaska. So he devoted himself to looking for new ways of using his resources and opened up a building business. Clearly uncomfortable in the tranquility of country life, which was a bit too slow for his liking, in 1911 he left for Tripoli where the Libyan question had broken out in the context of the Italo-Turkish war, and would lead to the annexation, as part of Italy, of the Tripolitania and Cirenaica regions. He stayed in Libya for a few months, and, while there was probably busy touting for possible orders for work, but without much success. In the meantime, his family continued to grow. Three more daughters were born: Alice Elena in 1910, Lidia Pierina in 1911 and Ester Maria in 1914.

Although Giovanni was more inclined to earn money by farming, he wasted no time in buying woods on Polè and Bolenghina, on the hill of Monfenera, which he transformed into agricultural land. The family blossomed: the first child was a son, Francesco, and then four daughters were born, two years apart from each other: Teresa Regina, called Teresina or Resi, in 1909, Marianna Elisabetta, called Bettina, in 1911, Margherita, called Rita, in 1913, and lastly Regina Italia, called Reginetta, in 1915. Jack was good father, and a passionate man: Rosina would describe his character well, remembering with a smile the time that he forced the coachman to stop so that he could promptly express his burning love for her.

In 1912 more good news came from Brazil: not only was the family growing fast before his eyes, but Gaspare, after adding his savings to the money he had received from his brothers had managed to buy a building that he planned to transform into a hotel in Guaporè. This venture, the natural evolution of the original hay-house, was named Hotel Roma and

Gaspare was congratulated by his brothers Giovanni and Francesco in the letter that they sent on April 1 of that year, which went as follows:

> We are pleased to hear about the acquisition that you have made and if, as you say, it is a healthy house, in a good position for installing an activity, with good surrounding land, you have done very well to have bought it thus ensuring a future for yourself and your family.

Replying to the request for financial aid that Gaspare must have made in his letter, they went on to say:

> We are prepared to do it on the next Saint Martin's day, when the tenant using our lands will pay us, as usual, the full sum stipulated, as long as the sum that you ask for in your next letter is not too large since, to tell you the truth, it is expensive to live decently in Italy, where the cost of food is still rising, especially now, due to the war with Tripoli — everything is sold at a higher price. And of the two that we once were, there are now twelve of us, that is to say, our wives, five girls and a boy, and two servants.

And with a little delay, in mid April 1913, Giovanni sent the fat sum of five hundred lire to Gaspare to help him sort out his hotel as well as possible. Around 1914 Giovanni and Francesco were still together in running a firm, called the Fornace Laterizi Sistema Hoffman F.lli Dalla Costa & C. [Brick and Tile Kiln — Dalla Costa Brothers & Co.), as shown on the headed paper used for a letter that they sent Giacomo at the end of 1913.

## Notes

1 From the recollections garnered from Giovanna Franca Dalla Costa during her talks with her aunt, Rita, it is known that Francesco had settled with his family in Pederobba in the large house adjacent to the Church Square that had been bought by his brother, Giovanni.

2 Recollections of Francesco Bonfiglioli.

3 Gaspare had ten children: Elisabetha, Lidia and Gentile, who never married, Rosalia, Aurelio, Francesco, Genoveva, Osorio Giacomo, Germano Gia-

como, and Ervino. At least, these are the children mentioned by Marly, Osorio Giacomo's daughter. The Parish of Guaporé was razed by a fire in 1930, and its old municipal hall was also destroyed, so it was not possible to search any archives to see whether there had been other offspring that had died soon after their birth, a relatively common occurrence in those times.

4 Letter sent from Pederobba.

5 Letter from Giovanni and Rosina to Gaspare, Pederobba, April 15, 1913.

6 Letter from Giovanni and Francesco Dalla Costa, Pederobba, without a date (probably December 1913).

# 28
# Reunion with Giacomo, and the War

Giovanni and Francesco were unexpectedly found by their brother Giacomo, who had, till then, been something of a mystery. It was known that he lived in France, but that was about all. There was simply no more news, either of him or his address. But then Giacomo learned from a friend of his who had been back to visit to Pederobba that some Dalla Costa family members were living there, and that they owned a factory producing building materials. Giacomo was overjoyed,[1] since he had almost given up hope of ever meeting his long lost brothers again. So he wrote to have the family members' news, just as the gathering storm clouds heralding the imminent outbreak of the First World War were causing him great apprehension, not only for himself but also for his family.

Giacomo now had a wife and two children; the elder one was called Clotilde, and the younger, Blanche. On 8 October 1902, he had married Marie Pauline Granet;[2] twelve years his junior, whom he had met while working in the French Midi, in the iron ore mine of Chateaudouble,[3] no more than twelve kilometers from Figanières, her birthplace. Then he had left the mine to do building work, but soon after was forced to leave because he had eczema to his hands, which meant that he could not touch water or cement. He was currently working as a carpenter and wood craftsman.

On being invited by his brothers to go visit them in Pederobba, he lost no time, as seen in a letter from Francesco to Gaspare, on 12 February 1914, commenting on the latter's arrival from France:

> Now we can give you some news. This morning, on the first bus, our brother Giacomo arrived here in Pederobba, or rather at our house; we had had news of him only two months before, after twenty years of hearing nothing at all. He is well and has been living in France

> for more than ten years, with his wife and two daughters, one aged ten and the other, five. He works as a carpenter and cask maker and lives a decent life. He had invited us brothers to pay him a visit, but since this was not possible, he turned up here himself. You should see how from the blond child that he was, he has now turned really dark. He is beginning to look like Giovanni, but much smaller and thinner.[4]

Giacomo returned indeed to the village of his birth where, together with his wife and daughters, he spent a couple of months. While in Pederobba, he was saddened to see those handsome copper pans that his parents were forced to sell on leaving Italy;[5] he saw them inside the homes of the families he visited. But at least his visits made him realize that this was no longer any place for him. There was no demand for his work as a carpenter and now, after such a long time away, he felt a bit like a stranger in his own land.

So Giacomo, whose brothers gave him presents — a beautiful suit, a coat and some gold trinkets[6] — said goodbye, particularly fondly to little Cesco, Giovanni's son, with whom he had a special bond, and made his return journey to Figanières, where his father in law and sister in law were still living following the death of the mother of his wife, Marie. But come what may he wanted to leave France as soon as possible. The atmosphere was increasingly sinister, pregnant with the threat of imminent war. He had become a French citizen, and could be called to arms right up to the age of fifty — and he was forty-four.

Mainly because of this, the idea of emigrating pervaded his future plans. And where could he disembark with the initial possibility of being helped out if not in Brazil? After all, he had heard that his brother Gaspare was doing well there with his new hotel. On June 6, 1914, a few days before the assassination of Francis Ferdinand of Habsburg, which provided the initial pretext for the outbreak of hostilities, Giacomo sent this heartfelt appeal to Gaspare:

> Write about the fates of our poor parents, sisters and brothers and sisters in law.[7] Tell me whether or not you are faring well in that country, if it is large or small, how many inhabitants it has. Give me news of your family. For myself, I have always worked in Europe and I mar-

> ried in France, in Figanières. I have been married for twelve years now, have two daughters, and I visited my brothers in Italy last spring. I work as a carpenter and casket maker. Tell me if in that country there is work for these crafts, let me know if a good day's pay can be earned there because here, in these countries, work is going very badly. If work were to go well in that country, if one were to be able to earn enough to live or if one could set up a fish trade, or open an inn or anything else, let me know since I would decide to come and see you.[8]

And so it was, and things probably would have turned out better if the other brothers had followed this example, too. In late December 1914, Giacomo and his family boarded the steamship, the Duca di Genova, which was bound for Brazil.[9] Giacomo was deserting France, which was at war with the central empires, and leaving his brothers behind in circumstances that are clearly described in a letter that Giovanni and Rosina sent to Gaspare on May 9 of the same year:

> We greatly fear that the terrible flagellation of war may come here to Pederobba, to our lovely Italy! They are now calling thc young to arms day by day, hour by hour, spouses and fathers of every class and, if we continue at this pace only women, the aged and babies will be left back here at home. All of us are crying, and sighing, and famine is coming fast. Let us hope that God will bring peace between populations and nations, otherwise things really do appear serious and terrifying![10]

Giovanni and Francesco were at no risk of having to serve in the army, which called up only men of a maximum age of forty, but the tragic events of the war were to have fateful consequences — also for the people in our story.

In the first two years, in fact, the war bloodied and destroyed the land to the east, the Carso and the mountains of Carnia, but after the defeat at Caporetto, it suddenly veered onto the Piave and, in the space of a day, Pederobba was in the front line. Rosa's brother, Raffaele, who was then serving as Mayor of the village, rushed to Giovanni's house to beg him to leave straight away and take his entire family with him. This was in November

1917, when the entire harvest had just been gathered, and the house was stocked to the brim, but everything was left behind when the family left, within hours of being warned. Giovanni did all he could to hide his effects, papers and photographs, in the bottom of a large vat under a pile of murk. But he put the more precious belongings in a small chest, which he coated with pitch so as to make it waterproof, and buried in the orchard in front of the kitchen, whilst trying to ensure that no-one saw him doing this.

Soon after taking these precautions, he left with his family members, in a carriage with a farm cart that was towed behind it holding little more than some mattresses. Little Francesco wanted to bring his new bicycle with him, too, no matter what. But, strapped between the carriage and cart, at the first bend the inevitable happened: it was crushed. And much more was to be crushed in those terrible times.

As evacuees in Pavia, in Piazza Borromeo 1, Giovanni and his family soon realized that they were not going to stay there for eight days as they originally thought, but for several months. Giovanni, who still had guts in spite of his years, went back on his own to Pederobba to try and retrieve some things that he really cared about. He returned at night, and on foot from Castelfranco, walked into the heart of the battle, lost his shoes which became entangled in the barbed wire, but managed to reach his home. This enterprise, however, only ended up with him seeing his home requisitioned, occupied by Italian soldiers, and in a shambles. Back to Pavia he went, heart in his heels, only to learn of the terrible suffering of little Resi: aged nine, she had died of the Spanish 'flu epidemic.

When the war was over, the return home was devastating: all that Giovanni had built thanks to the Alaskan gold was now rubble or had disappeared. Almost nothing remained of his large house but the indestructible four load-bearing walls: on entering the building, the family had the open sky above their heads because the roof had caved in. There was no trace of his carefully hidden precious belongings: the small trunk he had buried had been found, completely emptied out, and cast aside.[11] This was a disaster. But worse was to come. Giovanni found out that the gold coins deposited at the Società Operaia di Mutuo Soccorso [Workers' Mutual Aid Society] had been confiscated by the retreating Austrian army. He did all he could to get them back but, on reaching Fortezza, he lost all traces of his patrimony.[12] It was the final insult in the tragic fate of Giovanni, who had been so "lucky" in finding gold, and so unlucky in losing it, and he lost it only because, rather than depositing it in a bank on the right side of the river Piave, he deposited it in one on the left side, which had been conquered by the Austrians.

## Notes

1 Gilberte Arnoux describes that moment in the letter sent from Figanières on November 13, 2004 to Mariarosa Dalla Costa; in it she relates the statements made by her mother, Clotilde who, at the age of ten, saw her father Giacomo burst into tears, and exclaim to his wife: "Marie, Marie, I think I have found my brothers again!"

2 Dates traced in the marriage certificate issued by the Municipality of Figanières, Département du Var, Arrondissement de Draguignan.

3 The mine was opened by a German company in the eighteenth century and was closed in 1914, when it employed about three hundred workers, many of whom were Italians and Poles.

4 Letter sent from Pederobba.

5 Gilberte Arnoux, in her letter from Figanierès of December 30, 2004 to Mariarosa Dalla Costa, relates that her grandfather said to his wife: "Just look at that, Marie, all those copper pots and pans belonged to my mother."

6 It was probably then that Giacomo received the first nugget extracted by the brothers in Alaska; it was made into a pendant, the decoration on it showing a gold digger's pick and shovel. This piece of jewelry now belongs to Ivan Luiz Dalla Costa, son of Germano Giacomo, son of Gaspare. Giacomo took it with him when he went to Brazil so that he could give it to his brother, probably on behalf of Jack and Frank.

7 Neither of the parents, Luigi and Teresa, would have been alive then. Regarding the youngest sister, Elisabetta, on the other hand, the recollections of the family do not go beyond a vague version according to which she was "lost" in Argentina. No more news or official documentation regarding the other sister, Maria Anna, was forthcoming after the birth of the third daughter that she had conceived with her husband, Fortunato Viviani; from the text of the same letter we can deduce that she also emigrated with her family to Brazil. But there is no information available as to exactly where, or when. The only time that Maria Anna appears to be mentioned in the materials in our possession is in the letter sent from Pederobba on April 1, 1912 from Francesco and Giovanni Dalla Costa to their brother Gaspare at Guaporè in Brazil: "A daughter of poor Marianna wrote to us, and we will reply in a few days.".However the adjective "poor," used in this letter also to describe the parents, the sisters and the in-laws, suggests that they had died or had suffered an unhappy fate.

8 Letter sent from Figanières.

9 Postcard showing the same steamship, the Duca di Genova, sent by Marie to her father, Xavier Granet, when she had disembarked at Rio de Janeiro on January 19, 1915. On eventually reaching the brother Gaspare at Guaporè, Giacomo realized there were better work opportunities at Porto Alegre and

so he moved there with his family. Here his wife Marie, who was particularly enterprising, not only opened a grocery and refreshment store where she also sold liqueurs that she had learnt to make with herbs, but she was also quite successful as a French teacher. She had, moreover, learned to make paper flowers, and this skill earned her some extra money. From the letters that Marie sent to her father and sister who stayed in Figanières on February 6, March 25 and July 11 of 1917, January 1, May 18 and July 20 of 1918, September 13 and March 7 of 1920, it seems that Giacomo's family had managed overall to secure itself a pleasant and satisfying life-style, notwithstanding the constant risk of smallpox in the area in 1917, and the incurable skin condition of the hands that Giacomo suffered from; despite this he found a job and helped Marie in the shop. Food was plentiful, nature was rich and abundant. The large pine-nuts were a good substitute for chestnuts, and bananas and coconuts were plentiful. Then there were good sweet potatoes, and Marie promised herself that she would take some of their seeds back to France. The price of meat was reasonable and rents were affordable even if somewhat high. The anxiety that Marie expressed in these letters, obtained thanks to Gilberte Arnoux, were rather linked to the miserable condition of her parents in France during this time of war, to the constant worsening of her father's health, and Marie's awareness of the risk that the mayor might confiscate her family's land to settle the long-standing debt incurred when her mother was in the hospice, in which she had died. Starting from October 1917, when Brazil joined the allies in the war against Germany, Marie was given further cause for concern: there were restrictions due to the state of siege, marked increases in the price of food and she could no longer effectively help her parents in France by sending them essential goods because now the ships hardly ever made transportations for private individuals since the voyage was too dangerous due to the war being waged in the seas by the Germans; any free cargo vessels were now being used above all to transport food and raw materials to the allied armies in Europe. Marie only managed to send tiny packages weighing 350 grams compared with the parcels weighing 3 to 10 kilograms that had been allowed in normal times and, moreover, the postal charges had risen so much that they were higher than the value of the goods (foodstuffs like sugar and coffee) to be shipped. To add to her concern, there had been a plague of locusts that had devastated the area.

10 Letter sent from Pederobba.

11 With this robbery the photographs and documents concerning the Alaskan period also disappeared. Most of the photographs used for the present book were provided by the descendants of Frank Costa or by the brothers of Jack who emigrated to France and Brazil. Today the small wooden trunk, which has been restored and polished, is part of the family furniture.

12 Recollections of Jack's grandson, Giovanni (Gianni) Dalla Costa.

# 29
# A Bitter Tale

The postwar period was almost unbearable for the Dalla Costa brothers, who had constantly striven to affirm themselves with exceptional courage, willpower and determination. Their frame of mind in these difficult times is shown in the letters that they sent in that period to Gaspare, in one Giovanni and Rosina say:

> You know that like all the Veneto people we, too, in November of 1917 were under enemy bullets and had to abandon our homes, lands and harvests to migrate like poor pilgrims with no idea of where to take shelter with five children all under the age of ten. The terrible war forced us to leave everything we had in only one hour. We left without any clothes or money. On reaching, through sheer chance, the city of Pavia, apart from the discomfort of the journey, we encountered only poverty, misery, pain, humiliation and misfortune. Suffering and tears and damp air; and while, for more than a year and a half, we endured all this, we also suffered a terrible loss: the death of our eldest daughter, the one called Teresina after our deceased mother. After the longed-for armistice we returned to Pederobba where we found nothing but four collapsing walls, without any of those things that are necessary for life. Because the government is poor, it has not compensated us for even a part of the damage inflicted, you can only imagine therefore the difficulties we have in setting up our house again, even with the bare necessities for seven people. You should know that after our return we ran up a debt of 40,000 lire with the bank and this was done so the family could settle in, and for this we pay nothing less than seven and a half per cent (inter-

> est) a year. So calculate that the interest amounts to more than three thousand lire, such a steep sum that we will be forced to sell a part of our houses or lands. I assure you that I am in a sea of thought.

The above moving account is complemented by the one given by Francesco and Maria:

We have endured and suffered so much. Once accustomed to having enough comfort in the tranquility of our lovely home surrounded by peace and goodness and heavenly harmony, we had to face pain, deprivation, tears and disillusion. And even worse: being bullied and derided by cruel selfish speculators who also abound in our Italy. Who has been more unlucky than us from the Veneto? Who has suffered more than us because of the war? Except for the soldiers on the front, no-one has experienced so much pain and made so many sacrifices for their motherland. After the long-awaited armistice we left, or rather we left our hovel behind, intending to returning to our sacred land that had been martyred by enemy bullets in the hope of finding at least a room to put a blanket in. My God, what a disaster! It had been destroyed, devastated, and part of it had fallen. We found there was nothing left. All the bounty of God that had been left two years before, everything had disappeared and had been destroyed by vandals — soldiers and even civilians. Many who, before the retreat, had nothing to their name are now well stocked up with the belongings of others, whereas we and so many others who worked hard to earn a position ended up worse off. After the armistice, then, tired of still being emigrants in the world in which we had suffered so much, we came back to Montebelluna where we looked for two rooms to rent for six months, since luckily the village had remained intact although close to the famous Montello. But there, too, we had to suffer, deprived of everything, and having to start to buy all and sundry, from spoons to the bare necessities, without means yet at current prices. Damages for the war had not yet been paid, except to some lucky people who had known how to manage things well. Now, by the grace of God we are in our own house, which has been rebuilt by us, and we hope that soon we will be paid by the government itself. Of course it is now empty and nothing is in it, whereas before it was well furnished. Let us be patient. May the Good Lord keep us in good health and what will be will be. Anything can be borne except the loss of a person dear to us. Giovanni, for example, lost his eldest daughter in Pavia; ten years of age, she suffered from a merciless disease. Imagine his sorrow. We, on the other hand, saw our business affairs going from bad to worse, but we have always been healthy, and for this we praise and thank our Father in Heaven, and ask nothing from Him but the precious gem of health for all of us. Life here is difficult with prices these days, where goods and items of every sort cost twenty and thirty times more than they did once. It is enough to drive one mad. A pair of oxen cost 6,000, a horse from 3,000 to 20,000. Butter costs twenty lire a kilo, and so does cheese; milk, one hundred twenty a liter, meat ten or twelve a kilo, dried beans 35 and so on, shoes, 70, 80, 100 lire a pair, clothes 80 to a hundred lire a meter, and umbrellas 50 lire each. These things seem unbelievable. We hope that things will change, otherwise we cannot live.

## Notes

1 Letter sent from Pederobba on January 9, 1924.

2 Letter sent from Pederobba, without a date.

# 30
# The End

It was all about starting again, as Lidia, Francesco's daughter, remembers her father saying. And indeed, with the rise of Fascism, Francesco had the opportunity to do some of the jobs that were in enormous demand thanks to the public works policy of Mussolini.

On moving to Castelfranco, in the interests of his children's schooling, he started up a building and construction company in which he also found a way of employing his eighteen-year old nephew, Francesco, for a period of time. During the twenties, he worked in Calabria and Sicily. In the Veneto he carried out broadening and adjustment interventions on the banks of the Brentella canal along its entire course. Together with others, Francesco's company was also involved in the drainage of lands surrounding Meolo in the province of Venice. Thanks to these activities, he could afford to design and build a new house in Treviso, which he decided to call Villa Fairbanks, where he and his family settled at the close of 1928. The pretty villa, still in perfect condition, stands on the corner between Via Nazario Sauro and Via Luigi Pastro in the green and peaceful setting of the "garden city" of Treviso. The initials of the builder, FDC, in gilt letters, still stand out on the window panes.

When the war ended, Giacomo's main reason for continuing to stay in Porto Alegre vanished. There, in post-war Brazil, where the war had also led to restrictions and high increases in prices, the family had resumed its peaceful, calm and prosperous normal life. The only really pressing reason for the family to leave again for France was the recent news that Marie's father's health had deteriorated. So, on 28 March 1920, the family left Porto Alegre and, on 5 April, boarded a ship at Rio de Janeiro, bound for Genoa; in early May they at last reached Figanières.[1] Here they used up the savings accumulated on the other side of the ocean not only to salvage the family land by paying off the debt contracted for the hospice in which Marie's mother had been kept until her death, but also to secure the land for her final burial, and to set up a drafts workshop for carpentry where Giacomo could continue to work.

And Giovanni? Giovanni had embarked on no new enterprises. He continued to manage the few properties left to him, above all cultivating vineyards, and by processing grapes in the large cellars under his home, he produced wine that he sold in the surrounding areas. All in all he continued to lead the life best suited to him: his house was always open to friends, and his generous social nature, his need to sing in company, meant that he often announced to his wife, in his dialect: "Rosina, hey this evening the lads are staying to eat with me." And, almost every evening, ten or eleven of them were given food and drink.

Although in his life Giovanni had known riches, he continued to identify with the rural world in which he was born. He kept the maximum reserve and respect for peasants and miners. It is significant that, although he had a good overcoat, he never wore it because he did not want to "look like a gentleman," preferring to wear a simple jacket. His only quirks were the gold chain and watch. He never parted from them, not even when doing heavy work, such as stacking hay.[2] There was one particular song that he really loved, and he often persuaded Giustina,[3] a family servant with a good voice, to sing it. The words in the song, a tribute to love and generosity, went something like this: "Oh worthy people who love the needy, take care of the poor miner, and your reward will come from the Lord."[4]

The memories of Alaska were always vivid within him … one day, sometime around 1922, he set off with his horse and carriage, taking his wife and fifteen year-old son with him, for an eight-day trip to search for an old gold-digger friend whom he had met in Alaska. Having heard about his friend's return to Borca di Cadore; he searched for him high and low, but in vain.[5]

Every evening Giovanni sat on the balcony of his home on the landing that looked out onto the street. Sometimes, on finishing his rounds, Doctor Calvi dropped by, and readily accepted his friend's invitation to stop and have a glass of wine. During these peaceful meetings between friends, Giovanni would often tell stories of his life, and Calvi would listen with interest to another piece of the "Alaska story." Overall the old lion passed these years peacefully. With his family around him by the fire-place in the kitchen that looked out on the garden, he never boasted about his adventures to his children[6] although he did not fail to remind them of the sweat of the brow that their wellbeing had cost him. He would tell Francesco, in particular, about this, for he wanted to teach him to be aware and circumspect; in Giovanni's view, this son of his was a little too clothes conscious for his liking. And when he once saw him wear silk-lined knee-socks, he exclaimed: "Oh, yes, your knees are covered in silk whereas mine were torn on stones." On other occasions he gave his loved ones some pearls of wisdom, like the time he spoke to Rita, who had heard from the nuns that the elect were few and far between, and that the others were excluded; "But will the people who do not believe in God go to Hell?", she asked. "No," replied Jack "you should know that the world is like a circle with a bull's eye in the middle. You can get to the bull's eye from all parts, and along many paths, but you must always go in a straight line to hit it; if you don't, you will never get there."[7]

Although Rita was his favorite daughter, he had a particularly soft spot for Bettina, whose cordiality and industriousness in helping in the home he appreciated. But he was able to pay attention and demonstrate impartiality to his children when, in 1926, together with Rosina he decided to send Reginetta, who at the age of eleven demonstrated a particularly vibrant intelligence, to the Collegio Maria Immacolata of Conegliano [a Roman Catholic boarding school] so she could continue with her studies. It was a sacrifice to send her there, but both parents believed that it was important to help her in her studies.[8]

On 26 May 1928, the day of San Vettore Corona, Giovanni got into Doctor Calvi's car and together they went to Feltre, where a parish celebration was being held to commemorate the patron saint. It was a very special occasion in those times, worth remembering, they decided by taking a photograph of them both sitting on the car bonnet. It was to be the last snap.

On the following days Giovanni started complaining of acute earache. On 31 May, together with his daughter Reginetta, now thirteen years of age, he went to hospital, where the doctor advised him to go back home to bed. In the evening he got his son Francesco to go fetch his friend Lino Vercellino, whom he entrusted to take care of a few matters concerning the crops the following day. Then, when he started to feel seriously ill, Doctor Calvi was called in, and then another physician, and then another one from Onigo, who was renowned for being an exceptionally good practitioner. Lino Vercellino rushed to Onigo to get some oxygen.[9] But it was hopeless. That night Jack became delirious. His delirium was riddled with words uttered in English, some of them only comprehensible to Rosa. Probably in those hours, while he was dying, Giovanni

Dalla Costa relived the adventures of Jack Costa, saw once again the moose, the deer, the wolves, the large rivers, the roaring waterfalls, the forests, the mountains, the ice, the infinite white of the snow, and his dog Jocko; he probably re-lived the emotions felt in his search for gold, and the joy he felt that morning when he victoriously stuck his head out of the black mine shaft hole and shouted out to the world: "By Godda, I gotta de gold!"

Before six in the morning of the first day of June 1928, when his brother Francesco had arrived from Castelfranco, Giovanni Dalla Costa died of heart failure. He was exactly sixty years old. The man, one of the first to cross Alaska, who with his friends had founded the new city of Fairbanks, one who had lived a legendary life that had been as intense as it was strenuous, with triumphs and pain, was no more. Doctor Calvi spoke at the funeral and wrote the epitaph that can still be seen on the marble tombstone set in the wall girdling the small cemetery of Pederobba:

Giovanni Dalla Costa
humble mild good
tenacious worker
in hard tasks
in icy unexplored Alasha
gave his best years
then to his family
devoted his whole self
an early end took him away
leaving his dear ones in pain
in sadness — all the good mourned

The word 'Alasha' was neither an oversight nor a slip of the pen or chisel, but a mistake true and proper that speaks volumes, for it tells us all how far away this country was, even for an educated person like Doctor Calvi, who had spent so many evenings with Jack Costa, hearing him talk about his myriad adventures on ice; Alaska was for him, as it was for his countrymen, something magical, distant and intangible, something so elusive that one could not even focus properly on its name.

## Notes

1 Letter from Marie, wife of Giacomo, sent from Port Alegre to her father and sister at Figanières on March 7, 1920.

2 Recollections of Rita, garnered by Giovanna Franca Dalla Costa.

3 The person who sang was Giustina Bresolin, cousin of Giacomo Bresolin, the family coachman.

4 Recollections of Rita garnered by Giovanna Franca Dalla Costa.

5 L. Pellegrini, "Una sposina veneta in luna di miele tra i cercatori d'oro" [A Veneto Bride on Honeymoon in the Midst of Gold Diggers], in *Il Giornale,* December 11, 1990, p. 4.

6 "My father never really let us know how great his life had been. He did not attach that much importance to his achievements — he was very modest." Recollections of Francesco Dalla Costa (firstborn son of Giovanni) garnered by his daughter, Mariarosa.

7 Recollections of Rita garnered by Giovanna Franca Dalla Costa.

8 Recollections of Valeria Bresolin. Reginetta qualified as a school teacher in 1933.

9 Recollections of Francesco Dalla Costa, Giovanni's firstborn son. According to the recollections of Reginetta, garnered by her daughter, Valeria, the death was sudden, happening after a dinner enjoyed with friends. Perhaps Reginetta's apparently contradictory story is so simple because neither she nor either of her other sisters, Bettina and Rita, were told to wake up while the father was dying. Rather, they were woken at five thirty in the morning to go to the usual church service that, by tradition, they had to attend for nine Fridays. Only when they got back home from church were they told that their father had died.

# 31
# Epilogue

If life caused Giovanni great joy and acute bitterness, at least death saved him from seeing the final dissolution of all that he had built through a lifetime of toil. His unexpected and untimely death occurred just one year before the 1929 crash, which led to sky-high interest rates on loans taken out after the war, and which forced his widow Rosa to sell all the remaining estate — first the land and then the house. Rosa died in 1955, but not before having the satisfaction of seeing her three daughters married, and her son Francesco establish a solid position for himself, and build up a family with his wife Maria Ghidelli and their three children, Mariarosa, Giovanna Franca and Giovanni.

Francesco continued to live in Treviso until 1937. Advised by a doctor to change climate because of his asthma, together with his daughter Lidia, he went to see Florence. But this city was too big and crowded for his liking. He went back to the station and asked where the next train was headed. To Pistoia, they replied. When he got off the train he liked the silence, the few people, and the large avenue. In 1938 he decided to sell villa Fairbanks and moved to Pistoia with his entire family. Here he deposited all his money in a bank. He would have liked to invest it in an estate in Livorno because, he said, "If the war comes, everything will be gobbled up again by devaluation." But he never managed to achieve this aim — his energy was leaving him, and he felt his life closing in on him. He died in 1940 after spending his last few years reading the newspaper every morning while seated at the smartest café in the avenue, having made the request that the words on his tombstone should read: Francesco Dalla Costa, industrialist.

Giacomo continued to live together with his family at Figanières after returning from Brazil, working on little plots of land that he owned, but also doing work as a carpenter, managing the bar in the town square and opening a grocery store where Marie could continue to sell, along with everything else, cold cuts of salami and prosciutto that her husband was so good at preparing using the pork from pigs on their farm. He died in 1942, following surgery for an ulcer.

Gaspare continued to live with his family at Guaporè. In the thirties, a huge fire destroyed the parish and the local town hall, making it impossible to trace his children's birth dates. During the Second World War he was forced to change the name of his "Hotel Roma" and decided to rechristen it "Hotel Gaspare Dalla Costa." In 1953 two events occurred within a short space of time: Gaspare died and a fire razed his hotel to the ground.

# Bibliography

AA. VV., *Un altro Veneto. Saggi e studi di Storia dell'emigrazione nei secoli XIX e XX* [Another Veneto. Essays and studies on the history of emigration in the nineteenth and twentieth centuries], Franzina Emilio (ed.), Francisci, Abano Terme (Padova), 1983.

Adams C. W., Adams, C. W. (Charles W.) 1877–1962. *A Cheechako Goes to Klondyke,* John D. Williams and Trust Epicenter Press, Fairbanks, Alaska, 2003.

*Affari sociali internazionali* [Social and international affairs], no.3, 1973.

*Alaska Sportsman,* September 1961.

Bernardi Ulderico, *A catàr fortuna. Storie venete d'Australia e del Brasile* [Searching for fortune. Veneto stories of Australia and Brazil], Neri Pozza, Vicenza, 1994

Brunello Piero, *Emigranti* [Emigrants], in Lanaro S. (ed.), *Il Veneto,* Giulio Einaudi, Turin, 1984.

Cole Terrence, E. T. Barnette. *The Strange Story of the Man Who Founded Fairbanks,* Alaska Northwest, Anchorage, Alaska, 1981.

*Corriere di Caracas* (Il), no. 50, December 26, 1990.

*Daily News Miner,* (Successor to *Fairbanks Daily News*), "1902–1983 Special Golden Days Edition," July 18, 1983.

Dalla Zuanna Giampiero, Rosina Alessandro and Rossi Fiorenzo (eds.), *Il Veneto. Storia della popolazione dalla caduta di Venezia a oggi* [The Veneto. History of its Population From the Fall of Venice to Present Times], Ricerche Marsilio, Fondazione Nord Est, Marsilio, Venice, 2004.

De Felice Renzo (ed.), *Cenni storici sulla emigrazione italiana nelle Americhe e in Australia* [Historical Outline of Italian Emigration to the Americas and

Australia], FrancoAngeli, Milan, 1979.

Devoto Fernando J. et al., *L'emigrazione italiana e la formazione dell'Uruguay moderno* [Italian Emigration and the Creation of Modern Uruguay], Fondazione Giovanni Agnelli, Turin, 1993.

*Domenica del Corriere,* year 67, no. 37, September 12, 1965.

*Domenica del Corriere,* year 67, no. 38, September 19, 1965.

Duroselle Jean Baptiste and Serra Enrico, *L'emigrazione italiana in Francia prima del 1914* [Italian Emigration to France Before 1914], FrancoAngeli, Milan, 1978.

Erler Giorgio, "L'emigrazione italiana nello Stato di Rio Grande do Sul (Brasile) tra il 1875 ed il 1914" [Italian Emigration to the State of Rio Grande do Sul (Brazil) From 1875 to 1914], in De Felice R. (ed.), *Cenni storici sulla emigrazione italiana nelle Americhe e in Australia* [Historical Aspects of Italian Emigration to the Americas and Australia], FrancoAngeli, Milan, 1979.

*Fairbanks Daily Times,* August 25, 1906.

*Fairbanks Daily Times,* August 30, 1906.

*Fairbanks Evening News,* August 30, 1906.

*Fairbanks Daily Times,* August 31, 1906.

*Fairbanks Evening News,* September 12, 1906.

*Fairbanks Daily News,* November 23, 1906.

Ferrari Giovanni Lino, "Le forme nervose della pellagra curate con acido nicotinico" [Nervous Forms of Pellagra Treated with Nicotinic Acid], *Minerva Medica,* 1938.

Ferrazza Daniele, "All'ultima pepita. Da Pederobba all'Alaska" ["To the Last Nugget. From Pederobba to Alaska"], in *Il Mattino di Padova,* July 21, 1996.

Frank Costa, in *The Tanana Gold Fields,* single issue, Fairbanks, 1903, Rasmuson Library Archives, Cecil Robe Collection, Alaska and Polar Regions Department, University of Alaska Fairbanks.

Franzina Emilio, *La grande emigrazione. L'esodo dei rurali dal Veneto durante il secolo XIX* [The Great Emigration: The Exodus of Country People From the Veneto During the Nineteenth Century], Marsilio, Venice, 1976.

Franzina Emilio, *Merica! Merica! Emigrazione e colonizzazione nelle lettere dei contadini veneti in America Latina 1876–1902* [Emigration and Colonization in the Letters of Veneto Peasants in Latin America, 1876–1902], Feltrinelli, Milan, 1979, 2nd ed., 1980.

Franzina Emilio, *La transizione dolce. Storie del Veneto tra '800 e '900* [The Soft Transition: Veneto Stories From 1800 to 1900], Cierre, Verona, 1990.

Franzina Emilio, *Storia dell'emigrazione veneta* [History of Veneto Emigration], Cierre, Verona, 1991.

Gambasin Angelo, *Parroci e contadini nel Veneto alla fine dell'Ottocento* [Parish Priests and Peasants in Veneto at the End of the Nineteenth Century], Edizioni di storia e letteratura [Editions on History and Literature], Rome, 1973.

*Gazzettino* (Il), October 30, 1980.

*Giornale* (Il), November 21, 1990.

*Globo* (Il), [Australian newspaper] year 17 — issues 2 and 3, January, 1975.

Greever William, *La corsa all'oro* [The Gold Rush], Il Saggiatore, Milano, 1966.

Hanot Alex, "The History of Felix Pedro," in *Hanot Papers,* Rasmuson Library Archives, Alaska and Polar Regions Department, University of Alaska Fairbanks.

*Hanot Papers,* Rasmuson Library Archives, Alaska and Polar Regions Department, University of Alaska Fairbanks.

Jack Costa, in *The Tanana Gold Fields,* single issue, Fairbanks, 1903, Rasmuson Library Archives, Cecil Robe Collection, Alaska and Polar Regions Department, University of Alaska Fairbanks.

Jeier Thomas, *La corsa all'oro (Alasca 1896)* [The Gold Rush (Alasca 1896], La Scuola, Brescia 1977.

Lanaro Silvio (a cura di), *Il Veneto* [The Veneto Region], Giulio Einaudi, Turin, 1984.

*La valigia dell'emigrante* [The Emigrant's Suitcase], Editrice La Valigia, Vicenza.

Lazzarini Antonio, *Campagne venete ed emigrazione di massa (1866–1900)* [Veneto Countryside and Mass Emigration, 18676–1900], Istituto per le ricerche di storia sociale e di storia religiosa [Institute for Research on Social and Religious History], Vicenza, 1981.

Lazzarini Antonio, *Fra tradizione e innovazione. Studi su agricoltura e società rurale nel Veneto dell'Ottocento* [Between Tradition and Innovation: Studies on Agriculture in the Rural Veneto Society of the Nineteenth Century], FrancoAngeli, Milano, 1998.

Magliaretta Leopoldo, "Alimentazione, casa, salute" [Diet, Home, Health], in Lanaro Silvio (ed.), *Il Veneto,* Giulio Einaudi Editore, Torino, 1984.

Magnaghi Russel M., *Alaskans in the Alaska-Yukon Directories and Gazetteers 1902–1923,* self published, 1989.

Meneghetti Casarin Francesca, *Treviso-Genova, andata e ritorno. Gli albori dell'emigrazione transoceanica e l'inchiesta dell'Ateneo di Treviso* [Treviso-Genoa Return: The Dawn of Transoceanic Emigration and the Survey Conducted by the University of Treviso], Benetton Foundation-Il Cardo, Treviso-Venice, 1990.

Missori Marco, "Le condizioni degli emigrati alla fine del XIX secolo in alcuni documenti delle autorità marittime" [The Conditions of Emigrants at the End of the Nineteenth Century in Selected Naval Authority Documents], in *Affari sociali internazionali,* no. 3, 1973.

Morgan Lael, *Good Time Girls of the Alaska Yukon Gold Rush,* Epicenter Press, Fairbanks, Alaska, 1998.

Murray Jean, "Pedro's luck shone brightly on Happy Jack," in *Daily News Miner,* (Successor to *Fairbanks Daily News*), "1902–1983 Special Golden Days Edition," July 18, 1983.

*Notte (La)* [Night (The)], August 11, 1982.

Norris Frank, *North to Alaska: An Overview of Immigrants to Alaska, 1867–1945,* Alaska Historical Commission Studies in History, no. 121, June 1984.

Oddone Juan Antonio, "La politica e le immagini dell'immigrazione italiana in Uruguay, 1830–1930" [Policy and Pictures of Italian Immigration to Uruguay, 1830–1900], in Fernando J. Devoto et al., *L'emigrazione italiana e la formazione dell'Uruguay moderno* [Italian Emigration and the Formation of Modern Uruguay], Fondazione Giovanni Agnelli, Turin, 1993.

*Oggi,* year XXX, no. 50, December 11, 1974.

*Oggi,* year XXX, no 51, December 18, 1974.

Pellegrini Lino, "La febbre dell'oro" [The Gold Rush], in *Domenica del Corriere,* year 67, no. 37, September 12, 1965.

Pellegrini Lino, "Il modenese che vinse la corsa dell'oro" [The Man from Modena Who Won the Gold Rush], in *Domenica del Corriere,* year 67, no. 38, September 19, 1965.

Pellegrini Lino, "Luna di miele tra i lupi" [Honeymoon Amongst Wolves], in *Oggi,* year XXX, no. 50, December 11, 1974.

Pellegrini Lino,"Un orso lo salvò dal gelo" ["A Bear Saved Him from the Ice"] in *Oggi,* year XXX, no 51, December 18, 1974.

Pellegrini Lino, "Dopo anni di avventurose ricerche i fratelli Giovanni e Francesco Dalla Costa scoprirono un grosso filone aurifero" ["After Years of Adventurous Searching, Brothers Giovanni and Francesco Dalla Costa Discover a Big Gold Vein"] in *Il Globo* (Australian newspaper) year XVII — nos. 2 and 3, January 1975.

Pellegrini Lino, "Da solo nell'inferno buio" ["Alone in the Dark Inferno"], in *Il Gazzettino,* October 30, 1980.

Pellegrini Lino, "Pepite, lupi e...delitti" ["Nuggets, Wolves and ... Crimes"], in *La Notte,* August 11, 1982.

Pellegrini Lino, "Gli italiani e la febbre dell'oro," in *Storia illustrata,* no. 322, September 1984.

Pellegrini Lino, "Un trevigiano 'aprì' la corsa all'oro nel favoloso Yukon"

["Man from Treviso 'Opens Up' the Gold Rush in the Fabulous Yukon"], in *Il Giornale,* November 21, 1990.

Pellegrini Lino: "Un trevigiano 'aprì' la corsa all'oro nel favoloso Yukon" ["Man from Treviso 'Opens Up' the Gold Rush in the Fabulous Yukon"], in *Il Corriere di Caracas,* no. 50, December 26, 1990.

Pellegrini Lino, "Un'Alaska che ignoriamo. Viaggio nella 'febbre dell'oro,' cui partecipò l'Italia," ["An Alaska That We Ignore: Journey in the 'Gold Rush' That Italy Took Part In"], in *Qui Touring,* year XXVI, no. 3, March 1996.

"Pergamene antiche dei nobili trevigiani conti di Onigo (sec. XIII–XVIII)," Vol. II, *La prima dominazione veneziana (1339–1365), Pergamene 117–381,* [Antique Parchments of the Noble Treviso Counts of Onigo (Thirteenth to Eighteenth Centuries), Vol II, The First Venetian Domination (1339–1365) Parchments 117–381], Cataloguing by Gabriele Farronato, Antiga Edizioni, Cornuda (Treviso), 2001.

*Qui Touring,* year XXVI, no. 3, March,1996.

Sereni Emilio, I*l capitalismo nelle campagne (1860–1900)* [Capitalism in the countryside], Einaudi, Torino, 1956.

*Storia illustrata* [Illustrated history], no. 322, September, 1984.

*The Tanana Gold Fields,* single issue, Fairbanks, 1903, Rasmuson Library Archives, Cecil Robe Collection, Alaska and Polar Regions Department, University of Alaska Fairbanks.

*Twelfth Census of the United States. Census of Alaska.*

University of Alaska Students' Newspaper (*Farthest North Collegian*), October 1935.

Wickersham James, *Old Yukon: Tales — Trails — Trials,* Washington Law Book, Washington, D. C., 1938.

Zannini Andrea and Gazzi Daniele, *Contadini, emigranti, "colonos." Tra le pre-alpi venete e il Brasile meridionale: storia e demografia, 1780–1910,* 2 tomi [Peasants, Emigrants, "Colonos": From the Veneto Pre-Alps to Southern Brazil: History and Demography, 1780–1910, two volumes], Benetton Foundation Studies and Research, editorial coordination, Patrizia Boschiero, Canova, Treviso, 2003.

## Archives, libraries and other consulting bodies

Archivio di Stato di Treviso [State Archive of Treviso]

Archivio di Stato di Padova [State Archive of Padua]

Archivio parrocchiale di Pederobba (Treviso) [Parish Archive of Pederobba (Treviso)]

Archivio comunale di Pederobba (Treviso) [Local authority archive of Pederobba (Treviso)]

Civica Biblioteca – Archivi Storici "Angelo Mai" – Bergamo [Civic Library – "Angelo Mai" Historical Archives — Bergamo]

Fondazione Ansaldo di Genova [The Ansaldo Foundation of Genoa]

Memorial do imigrante, San. Paolo, Brasile (http://www.memorialdoimigrante.sp.gov.br)

Rasmuson Library Archives, Cecil Robe Collection, Alaska and Polar Regions Department, University of Alaska Fairbanks

Elmer E. Rasmuson Library, Alaska and Polar Regions Department, University of Alaska Fairbanks

Ellis Island Records (http://www.EllisIsland.org)

Mairie de Figanières, Département du Var, Arrondissement de Draguignan, France

Municipio do Guaporé, Rio Grande do Sul, Brasile

Library of Congress, Washington, USA

Biblioteca Ettore Anchieri della Facoltà di Scienze Politiche dell'Università di Padova [Ettore Anchieri Library of the Faculty of Political Science, University of Padua]

Biblioteca Universitaria Padova [University Library — Padua]

Conservatoria dei registri immobiliari di Treviso [Custodial office of the land registers of Treviso]